CORRELATES OF FERTILITY BEHAVIOUR

CORRELATES OF FERTILITY BEHAVIOUR

A Study of Rural Communities in Punjab and Haryana

Dr. (Mrs.) K.P. SINGH

CONCEPT PUBLISHING COMPANY, NEW DELHI

First Published 1986

(Mrs.) K.P. **Singh** (*b*. 1936-)

Published by

Naurang Rai
Concept Publishing Company
H-13, Bali Nagar
NEW DELHI - 110015 (India)

Printed in India by

Aay Kay Prints,
at Mohan Printing Corporation
238, Padam Nagar,
DELHI - 110007.

ACKNOWLEDGEMENTS

THE present study is an outcome of a project undertaken by the Population Research Centre. It gives me great pleasure to record my deep sense of gratitude to Prof. V.S. D'Souza, Director of the Centre, for permitting me to collaborate in this project and to write the report. I am highly indebted to him for his encouragement, advice and all the facilities which he extended during the course of this study.

Financial assistance for the publication of this study was provided by the Department of Sociology, Panjab University, Chandigarh, out of the publication grant under the UGC Special Assistance Programme. I am indeed very grateful to the Department for this gesture.

I also acknowledge my gratefulness to other participants in this project—namely, Mr. R.S. Goyal, Senior Research Officer, who with tireless zeal conducted and supervised the various plans of this work; Mr. Gurdarshan Singh, Mr. Swarn Singh, Ms. Rita Gawari and Ms. Reena Singh, who as Research Assistants cheerfully laboured at the tedious task of data collection and processing.

I also acknowledge the useful suggestions concerning the statistical aspects of the study offered by Shri Kashev Chander Kaistha.

My acknowledgements are also due to Mr. K.K. Kalia, and Mr. Sudesh Kumar for typing this manuscript.

(MRS.) K.P. SINGH

CONTENTS

1

INTRODUCTION

INDIA has at present about 15 per cent of the world's population and this 15 per cent lives on less than 2.5 per cent of land area of the world. The present population of the country according to 1981 census is 684 million and it is growing at the rate of 2.2 per cent per annum.

The steep rise in the country's population during the last 50 to 60 years due to the decline in the mortality rate has outstripped the decrease in the birth rate. The death rate came down from 42.6 in 1911 to 14.5 per thousand in 1974 while the birth rate could be brought down from 49.2 per thousand to only 34.5 per thousand during the corresponding period.

The death rate has been brought down quickly and with comparative ease by controlling the occurrence as well as reducing the severity of famines, epidemics, communicable diseases and other natural calamities through modern public health measures and control of environment, but no such effective measures are available for quick reduction in birth rate as it involves a radical change in the traditional attitudes of the people about measures to effectively curb the rapid increase in population.

As a result, the rapid increase in population is posing serious threat to the country's developmental efforts. Provision of food, housing, health, employment and educational facilities poses a formidable challenge to the country as the resources available are not sufficient to meet the rising demands of the increasing population.

Thus, under the present circumstances, when the death rate has been controlled easily because the measures to reduce it are exogenous in nature, the reduction in birth rate calls for planned and strenuous efforts on the part of the government and other voluntary organizations because it involves bringing about change in the life philosophy and life style of the people so that they accept the norm of small family and practise family planning as a way of life.

In pursuit of the objective of accelerated economic and social development, since the early 1950's, the Government of India has formulated an official policy with the object of reducing the rate of population growth. The programme to control fertility was accorded special priority during the late nineteen sixties and nineteen seventies, but in spite of that the programme has so far met only with marginal success and the target of reducing the birth rate from 40 to 25 per thousand population by 1981, has remained only a pious wish.

This marginal realization of the aims and objectives of the family planning programme, may be attributed to many inextricably interrelated cultural and socio-economic factors which are intrinsic to the Indian society. However, it has been noted that people belonging to different socio-economic and religious groups desire children for different reasons and have different attitudes towards family size. The present study, therefore, is concerned with some of the determinants of fertility behaviour in rural communities.

A variety of factors have been identified as being, directly or indirectly, responsible for differentials in fertility at different levels. The United Nations (1973) report observes, "within limits established by physiological factors, a multitude of economic, social and cultural factors are the ultimate determinants of fertility levels and the variations in different societies."

Early studies concerned with fertility differentials have mainly been concerned with understanding the differentials per se, rather than providing explanations for the phenomena observed. With increasing recognition of the problem of over-population, as pointed out by Ghosh (1981), and the appreciation of the need for fertility control through policy and programme, the emphasis now is to seek explanations for observed fertility differentials and subsequently to develop theories and models.

There are large number of studies available and most of them (Agarwala 1964; Chandrasekheran 1952; Davis 1951; Driver 1963; El-Badry 1967) have analysed these differentials according to socio-economic variables such as income, occupational and educational levels etc. The economists have focused their attention mostly on economic variables paying little thought to social and psychological variables; whereas, sociologists have accorded less importance to economic variables as they emphasize that cultural factors play an important role in determining the acceptance of a small family norm. Goldberg (1957) argued that explanations of differential fertility in relation to socio-economic status as they left a large gap between independent and dependent variables. Stycos (1963) and Davis also pointed out that one of the serious obstacles to the family planning programme was the lack of information on motivational and sociological aspects. Therefore, there is need to analyze determinants of fertility behaviour in a broader way that would cover socio-economic as well as sociological variables.

The present study, therefore, proposes to examine the present levels of fertility among different socio-economic and religious groups and investigate some of the correlates of fertility behaviour in the rural areas of Punjab and Haryana.

Studies of fertility levels and their differentials often seek to measure directly the impact of socio-economic factors on fertility. But as pointed out by Bongaarts (1978) substantial insights can be gained, if, in addition to the socio-economic factors influencing fertility, the specific mechanisms through which these factors operate are also identified. A more detailed analysis may show that educated women marry relatively late or use contraception more frequently, thus clarifying the relationship between education and fertility. It, therefore, becomes important to understand the biological and the behavioural factors through which socio-economic, cultural and other sociological variables operate. These are termed as intermediate variables.

These relationships have been recognised in the pioneering work of Davis and Blake (1956), Freedman (1967), Hill, Stycos and Back (1959) and Westoff and others (1961).

A conceptual scheme provided by Davis and Blake (1956) has given a classification of intermediate variables affecting fertility. They have divided these intermediate variables into three broad groups namely 'intercourse variables', 'conception variables' and 'gestation variables'. Intercourse variables are further sub-divided into six parts, conception variables into three parts and gestation variables into two parts respectively. This model envisages that any factor—environmental, socio-economic, biological, psychological or cultural which affects fertility must go through a series of 'intermediate variables''. Hill, Stycos and Back (1959) while emphasising the family characteristics devised a conceptual model in which antecedent variables, intervening variables and consequent variables have been incorporated. An important conceptual model has been proposed by this sociologist Freedman, (1967). According to him socio-economic structure is in mutual influence relationship with the environment and mortality rates and influences norms concerning family size and intermediate variables which finally determine the fertility behaviour. The former set of variables directly and through the latter two variables influence the intermediate variables and all these, in turn, influence fertility as is evident from the diagrammatic presentation of Freedman's model.

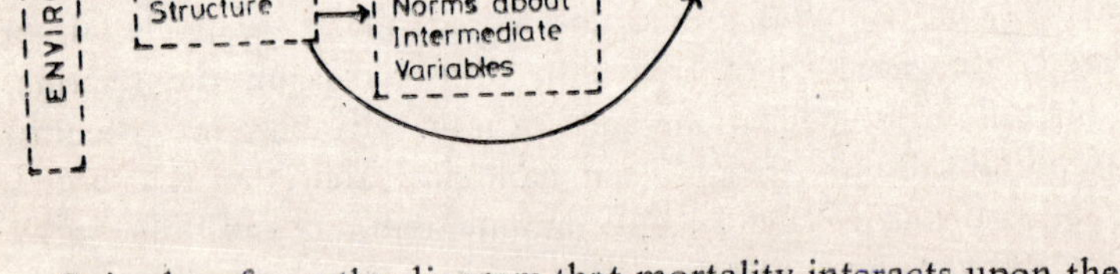

It is clear from the diagram that mortality interacts upon the socio-economic structure, which, in turn, effects norms about intermediate variables affecting fertility. Mortality and socio-economic structure are, therefore, the most important factors affecting fertility.

Making use of the above formulation as a starting point, the present study proposes to discuss the correlates of fertility behaviour under two broad headings, namely, individual variables and familial variables. Under the individual variables would be included demographic, socio-economic and other sociological variables, whereas, under the familial variables would be included the family structure, the relationship between husband and wife in terms of communication pattern and the reasons for begetting sons and daughters.

The study of the correlates of fertility behaviour in rural areas is of great significance because it can help in understanding the mechanisms of fertility decline, and, hence can be very helpful in the formulation of population policy as about 75 per cent of India's population lives in rural areas. It is also important from the point of view of the implementation of the family planning programme because it can assist in identifying high fertility groups which need to receive special attention at the hands of family programme field staff.

Universe of the Study

As mentioned above, the present study concerns itself with the determinants of fertility behaviour in the rural areas of Punjab and Haryana.

The states of Punjab and Haryana occupy 14th and 15th rank, respectively in terms of their population in the country's population. In Punjab (1979), the estimated annual growth rate of population is slightly lower than the national level, while in Haryana it is identical with the national growth rate. But the rural areas of these two states have slightly higher rates of population growth in comparison with the growth rate in rural areas for the country as a whole. Since in both the states more than three fourths of their population resides in rural areas, and as these rural areas are undergoing rapid socio-economic changes due to the green revolution, it was decided for the purpose of present study to focus attention on the rural areas of these two states.

Sampling Procedure

The study confines itself to three villages and these villages have been so chosen as to encompass as much diversity as possible. The villages are located in three different districts, two in Punjab and one in Haryana. Although the number of villages to be studied, the states and the districts in which they are located, were decided arbitrarily, yet the villages themselves were selected on the basis of four important criteria.

The criteria of selection are: (a) the villages to be selected should be large enough, say, with a population of over 4,000 so as to be capable of revealing a wide variety of change; (b) they should be outside the zone of influence of their nearest town and city. A distance of 10 km from the nearest town and 25 km from the nearest city was considered adequate to meet this requirement; (c) the population growth rates during the past three decades should be consistently high, or, consistently low as compared to the average growth rates of their respective districts and states and (d) the economy should be predominantly agriculural so that, at least, two-thirds of its workers should be engaged in the agricultural sector.

Thus two villages, one from Punjab (Ghagga) and the other from Haryana (Ugala) were chosen for their excessively high rates of growth of population during the 1951-61, 1961-71 and 1971-81 decades; and one village from Punjab (Ghudani Kalan) was chosen for its excessively low growth rates.

Sampling

The sample of 33 per cent of the households was drawn according to the procedures of simple random sampling. For the selection of household from every village, houselists prepared for 1981 census operations were used as the sampling frames for the present study. Since the unit of analysis was the household, all the married women residing in the household were included in the study. The sampling frame and the sample size for the three selected villages are given below :—

Village	District	State	No. of households constituting the sampling frame	Sample size of women
Ghagga	Patiala	Punjab	770	273
Ghudani Kalan	Ludhiana	Punjab	741	300
Ugala	Ambala	Haryana	598	235
Total			2109	808

Tools of Data Collection

A preliminary interview schedule was developed, prestested and revised and then used for data collection. The schedule consisted of three parts. In the first part was included general identification variables like age, education, occupation, income, landownership and family structure. The second part consisted of detailed fertility history of the family alongwith the history of infant and child deaths. The third part comprised of some of the sociological variables such as stratification variables, inter-spouse communication, religion and the level of religiosity.

Women were interviewed by the female investigators at their places of residence. During the course of data collection, the investigators stayed in the village, this facilitated establishment of better report with respondents and helped the investigators in getting full cooperation from the respondents.

SCHEME OF ANALYSIS

Fertility Variables

Two measures of reproductive behaviour, namely, number of live-births and desired ideal family size and, at places, the adoption of family planning practices, have been used as dependent variables and all other variables have been used as independent variables.

Independent Variables

A number of socio-demographic and sociological variables which do play an important role in determining fertility behaviour have been taken as independent variables. These variables have been divided into two parts and are given below:

(1) *Individual Variables*: These variables are discussed under three sub-headings. They are:—

(a) *Demographic variables* which include present age of the respondents, their age at marriage, age at the birth of the first child, duration of married life and the extent of infant and child mortality.

(b) *Socio-economic variables*: Under these variables are included educational level of the respondents and their husbands, their occupational status, family income, land-ownership, standard of living and housing conditions.

(c) *Sociological variables* include caste, religion and level of religiosity and modern and traditional value orientations.

(2) *Familial Variables*: Since the desire for a particular family size and adoption of family planning methods are influenced by family and the community to which one belongs these variables include the type of family structure, the husband wife communication, desire for more children, perceived advantages and disadvantages of large family and reasons for having sons and daughters.

The list of the factors is, however, not an exhaustive one as there are other demographic factors which influence fertility but those factors do not came under the purview of this present study.

Data Presentation

The qualitative data has been presented in the form of simple frequency tables with percentage distribution and in most cases in the form of two-way tables, according to the independent and dependent variables. Nevertheless, statistical tests, wherever required, have been worked out to determine the level of significance.

Preparation of the Indexes

In sociological research, it is not always possible to obtain information about a variable by a single question. Such information is usually obtained by a battery of questions and these questions are organized in the form of an index or scale. Then numerical values are assigned to these questions and by pooling such values, an index is constructed. In the present study, also, three different indexes were prepared to measure certain variables. These are (a) level of living, (b) Index of religiosity, and (c) index of the level of modernity. These indexes have been prepared to divide the sample population into three categories, namely, high, medium and low with regard to each of these variables.

Level of Living Index

Various measures concerning socio-economic status have been designed by taking into consideration education, occupational and income levels of the respondents, but not much attention has been paid to measure the level of living. However, in the absence of any standarized index as an indicator for the level of living especially in rural areas, a small index was devised to measure the level of living of the respondents. The items included in the index are (a) house type, (b) number of living rooms, (c) other housing amenities, (d) electricity, (e) water supply, (f) radio/T.V. and (g) means of conveyance. Every one of these items was assigned a score of one and the highest score for all the items was 7 and the lowest being 0. The scores were than divided into three categories:

1. 0,1,2 Scores represented low level of living.
2. 3,4 Scores represented middle level of living.
3. 5,6,7 Scores represented high level of living.

Index of Modernity

Since fertility behaviour is closely associated with the level of modernity and modern women have generally lower fertility level as compared with traditional women, it was decided to prepare an index of the level of modernity of the respondents. Therefore, a traditional modernity index consisting of nine statements was formulated.

For the purpose of identification of the level of modernity of the respondents, the responses made by each respondent were scored and then the scores for all the respondents were pooled. High score indicated modern outlook, whereas, low score indicated traditional outlook. The highest score was 9 and the lowest being 0. The total score was divided into the following three categories:

1. 0,1,2 Scores represented traditional respondents.
2. 3,4 Scores represented moderately modern respondents.
3. 5,6,7,8,9 Scores represented modern respondents.

The cutting points in the above index are not equally distributed because beyond score 4, the number of respondents who obtained a score of 5, 6 or 7 was rather small. Therefore, in order to have sufficient number of respondents, in the last category of modern respondents to enable us to make valid generalizations, last five values have been pooled into one category.

Index of Religiosity

Apart from one's belief in religion, an individual's commitment to religious activities also influences fertility behaviour. Therefore, an index of religiosity was prepared to examine the extent of respondent's commitment to religion. This index was again prepared in the same manner as the other two indexes. For this purpose, only four statements connected with the religious beliefs of the respondents were taken into account and the highest score was four, whereas, the lowest-score was zero. The scores were than divided into the following three categories:

Score zero represented highly religious respondents.
Score one represented moderately religious respondents.
Score 2, 3, 4 represented respondents who are least committed to religious activities.

Here again the scores have not been equally distributed because the frequency distribution of the respondents in score 2, 3 and 4 was rather small to permit any valid generalizations, hence they have been pooled into one category.

REFERENCES

1. Agarwala, S.N. 1964: "Social and Cultural Factors affecting Fertility in India", Population Review, 8.
2. Bangaarts, John 1978: "A Framework for Analyzing the proximate determinants of Fertility", Population and Development Review, 4:1.
3. Chandrasekharan, C. and George, M.V. 1952: "Mechanisms underlying the differences in Fertility Patterns of Bengalee women from three Socio-Economic groups", Milbank Memorial Fund Quarterly, 40.
4. Davis, Kingsley 1967: Population Policy: Will Current Programmes Succeed? Science, C1 viii.
5. 1951: The Population of India and Pakistan, Princeton, Princeton University Press.
6. Davis and Blake 1956: Social Structure and Fertility: An Analytical Framework in Economic Development and Cultural Change, Vol. 4.

7. Driver, E.D. 1963: Differential Fertility in Central India, Princeton, Princeton University Press.
8. El-Badry, M.A. 1967: "A Study of Differential Fertility in Bombay", Demography, 4.
9. Freedman, R. 1967: "Application of the Behavioural Science to Family Planning in Studies in Family Planning", 23 October.
10. Freedman, Ronald 1961-62: "The Sociology of Human Fertility: A Trend Report and Bibliography", Current Sociology, Vol. X/XI, No. 2.
11. Ghosh, B. 1981: "Policy Oriented Research into Determinants of Fertility", Demography India, Vols. 1 & 2.
12. Govt. of India: Sample Registration Bulletin, Vol. xiii, 2 (Vital Rates), December.
13. Hawthorn, Geofrey 1970: The Sociology of Fertility, London, Collier Macmillan.
14. Herr, D.G. 1966: "Economic Development and Fertility", Demography, 3.
15. Hill, R., Stycos, J.M. and Back K. 1959: The Family and Population Control, Chapel Hill, University of North Carolina Press.
16. Mahadevan, K. 1979: Sociology of Fertility: Determinants of Fertility Differentials in South India, New Delhi, Sterling Publishers.
17. Mitra, Asok 1978: India's Populations: Aspects of Quality and Control, Vols. I & II, New Delhi, Abhinav Publications.
18. Nam, Charles B. 1968: Population and Society, Boston, B. Miffin.
19. Potts, Malcolm and Peter, Selman 1979: Society and Fertility, London, Macdonald and Evans.
20. Stycos, J.M. 1963: "Obstacles to Programmes of Population Control—Facts and Fancies", Marriage and Family Living, 25.
21. Stockes, C.S. 1973: "Family Structure and Socio-Economic Differentials in Fertility", Population Studies, 27 (2).
22. United Nations 1973: The Determinants and Consequences of Population Trends, New York.
23. Yankey, D. 1969: The American Sociologist, No. 4, May.

2

SOCIO-ECONOMIC INFRASTRUCTURE OF THE STUDY VILLAGES

SINCE people belonging to different socio-economic and religious groups desire children for different reasons and have different attitudes towards family size, an understanding of the variables of socio-economic infrastructure would, therefore, serve a useful purpose in understanding differentials in their fertility behaviour. With a view to presenting a comparative picture, the infrastructural variables employed for the present study in the three villages are described under various headings.

Location and Population

As pointed out earlier the three villages were so chosen as to be relatively distant from the nearest town or the city and with large population than is usually found in villages. Ugala is at a distance of 14 km. from the nearest town, Shahabad, and 34 km from the nearest city Ambala (U.A.), Ghagga is at a distance of 19 km from the nearest town, Samana, and 45 km from the nearest city, Patiala, and Ghudani Kalan is at a distance of 20 km from the nearest town, Khanna, and 42 km from the nearest city, Ludhiana. These villages are connected with the towns nearest to them by road as well as by telephone.

According to the census of 1981, the population of Ugala, Ghagga and Ghudani Kalan were 4800, 5143 and 4380, respectively. Although all the villages are large, Ghagga is the largest and Ghudani Kalan is the smallest among them.

Land and Agriculture

Agriculture is the mainstay of the economy in all the three villages. The land under all uses in Ugala, Ghagga and Ghudani Kalan is 3511, 4477 and 3750 acres respectively, which shows marked variation; but the variation is much less in respect of cultivated land which is 3464, 3500 and 3419 acres, respectively. The per capita cultivable land in Ugala, Ghagga and Ghudani Kalan works out to be 0.72, 0.68 and 0.78 acre respectively.

Although the area under cultivation is more or less the same in all the villages, but the pattern of distribution of land varies. In Ugala and Ghudani Kalan about 48 per cent of the households do not own any land but this percentage is 60 for Ghagga. However, between Ugala and Ghudani Kalan the former is better off in terms of average land owned, taking into account either all the households or only the landowning households. Further, among these who own land, highest proportion of households owned one to 10 acres of land in village Ghudani Kalan, followed by village Ugala and Ghagga. On the other hand, the highest proportion of households in village Ugala owned more than 20 acres of land. Thus, on the whole Ugala occupies a pre-eminent position in terms of landownership and status of the households.

Since around fifty per cent of the households do not own any land many members of the households were engaged in different tasks. Table 2.1 gives the pattern of distribution of workers in nine industrial categories according to the estimates based on 33 per cent sample of the households.

Table 2.1 indicates that village Ghudani Kalan has the highest percentage of cultivators but the lowest percentage of agricultural workers. On the other hand, Ghagga has the lowest percentage of cultivators but the highest percentage of agricultural workers.

Thus, on the whole, agriculture is still important in all the three villages as in every village the single largest percentage of workers are engaged in agriculture. However, the other two major sectors which have absorbed workers are trade and commerce and other services. Ghagga has much higher percentage of workers engaged in trade and commerce.

An overall comparison between the landownership pattern and the pattern of distribution of workers in industrial categories shows that Ghagga has the smallest percentage of landowning households but the highest proportion of workers are engaged

TABLE 2.1

Percentage distribution of total workers in three villages according to the national industrial classification of workers for 1980

	Ugala	Ghagga	Ghudani Kalan
Cultivators	44.35	38.45	52.19
Agriculturral labours	22.87	32.92	17.32
Live stock/forestry/ others	—	—	—
Household industry	8.06	2.25	3.51
Manufacturing other than household industry	1.65	—	0.22
Construction	4.41	1.64	4.17
Trade and commerce	7.71	12.68	6.58
Transport/storage/ communication	0.28	0.82	0.22
Other services	12.67	11.24	15.79
Total number	1089	1467	1363

in trade and commerce. It is, thus, developing as trade centre. Between Ugala and Ghudani Kalan each of which has almost the same percentage of landowning households, Ghudani Kalan has a greater percentage of cultivators.

Facilities for Health and Family Welfare

Public medical facilities in the rural areas in Haryana and Punjab are provided through the Primary Health Centres (PHC), rural dispensaries, both allopathic and ayurvedic and their sub-centres. The PHCs are usually located at the Block headquarters and the allopathic and ayurvedic dispensaries cater to the needs of groups of villages simultaneously.

The public health facilities available in all the three villages are coordinated by PHCs located outside each village. The PHC responsible for Ugala is located at Mulan which is at a distance of 15 km and the nearest rural dispensary is in Barara, 7 km away. The only health facility located in the village is a rural sub-centre run by an ANM with the help of a 'dai'. A doctor from the rural dispensary visits it irregularly. The ANM is expected to conduct deliveries, provide maternity and child health (MCH) services, and locates the eligible couples for the family planning programme. A malaria worker also visits the village regularly. Besides the government health facilities, there are eight private medical practitioners but none of whom is a fully qualified doctor.

The PHC taking care of Ghagga is located at Satrana, 15 km away, and the nearest rural dispensary is at Patter, 11 km away. The village, however, has an ayurvedic dispensary headed by a vaid and a sub-centre of the rural dispensary. There is a malaria worker stationed in the village. Besides these facilities the village is served by seven private medical practitioners.

Relatively speaking, Ghudani Kalan is better provided with health care facilities. The PHC responsible for this village is located at Payal, barely 5 km away. There is a rural dispensary and an ayurvedic dispensary besides a sub-centre located in the village itself. The rural dispensary is headed by a fully qualified doctor who is assisted by an ANM and other paramedical staff. The ayurvedic dispensary is looked after by a vaid. A malaria worker also visits the village regularly. There are also seven medical practitioners in the village. Thus. in every respect Ghudani Kalan has superior medical facilities as compared with the other two villages.

Educational Facilities

In all these villages, there is provision of schools for studying upto matriculation standard. The schools are run by the governments. However, the high school at Ghudani Kalan has been in existence since 1959, while the high schools in the other two villages were established much later. Ugala and Ghagga had only primary schools to begin with, and they were upgraded to the high school level in 1961 and 1965, respectively. Ghudani

Kalan and Ghagga also have access to colleges which are situated at a distance of 2 km and 5 km from each of these two villages, respectively. On the whole, Ghudani Kalan has better educational facilities as compared with the other two villages.

Trade, Industry and Commerce

Village economy is based principally on agriculture. Consequently the development of each village is dependent upon agriculture. However, a higher order of growth would depend upon a sectoral transformation so that the non-agricultural sectors begin to contribute more and more towards its development. The growth of trade, commerce and industry in a village would, therefore, be important indicators of its economic development.

Large villages such as the ones we are dealing with, would naturally require a sufficient number of shops to cater to the daily shopping needs of the villagers. But there is considerable variation in the size and pattern of the marketing facilities available in these villages.

In Ghudani Kalan, the market is very small and the shops are scattered all over the village. There are in all 22 shops, almost all of them are small and all cater to the needs of the local population exclusively. For major purchases the villagers have to go to the town of Khanna.

The market at Ugala is of a modest size and the shops are somewhat scattered. There are in all 30 shops. By and large, as in Ghudani Kalan, the market in Ugala serves its residents only.

But the market at Ghagga is highly developed and serves as a shopping centre for the people of the surrounding villages besides its own residents. There are nearly a hundred shops most of which are situated along the main road near the bus stand quite close to the village.

Industry is not well developed in any one of these villages. However Ghagga has a fairly large rice mill employing 40 persons and Ugala has a crusher for sugarcane, which are operated seasonally.

An important cottage industry which is found in all the three villages is poultry farming. Ugala, Ghagga and Ghudani Kalan have three, four and twelve poultry farms, respectively.

All the three villages are well provided with banking facilities and every village has branch offices of two banks.

Administration

With the inception of village self-government after independence, Panchayats were established in the villages in 1952. Starting with five-members, the strength of the Panchayat members has increased with the growth of population. The latest elected Panchayats had nine members in Ugala and 10 members in Ghagga and Ghudani Kalan each. The annual income of the Panchayats in Ugala, Ghagga and Ghudani Kalan in 1980 was Rs. 30,000, Rs. 45,000 and Rs. 26,000 respectively.

In 1978 Ghagga was recognized by the Punjab Government as an urban centre, and its Panchayat was replaced by a Notified Area Committee (NAC). With the setting up of the NAC, Ghagga has better administrative facilities. It also has a police station whereas Ugala and Ghudani Kalan are served by the nearby police stations located at Barara and Payal, respectively.

An overall assessment of the socio-economic variables in the three villages indicates that Ghagga and Ghudani Kalan are more developed than Ugala. In this respect the two rapidly growing villages with regard to population, namely, Ugala and Ghagga, are further apart in terms of some of the indices of development. However, the greater development that has occurred in Ghagga and Ghudani Kalan is along different variables. Ghagga is more developed in respect of transportation and communication, administrative structure and trade and commerce; whereas Ghudani Kalan is more advanced agriculturally, educationally and in health services. In their development, Ghagga is acquiring urban characteristics whereas rural characteristics still prevail in Ghudani Kalan, however, with improved amenities of life. Ugala is, relatively, the least developed and lags behind these two villages in every respect.

3

AGE AT MARRIAGE AND FERTILITY

One of the important determinants of fertility is marriage. The ages at which men and women marry and also the extent to which marriages occur have an important bearing on fertility.

Age at marriage is a significant factor in determining fertility because it marks the beginning of the reproductive span of a woman's life and any shift in the age at marriage can postpone the beginning of the reproductive process. It also determines the total duration of marriage which influences the fertility behaviour. But age at marriage in itself is influenced by various sociological variables such as education and gainful employment of women which, in turn, influence the attitude and behaviour of women affecting, thereby, their attitude towards family size. Davis (1964) also observes that age at marriage is determined not so much by law as by the existing socio-economic conditions prevalent in the society.

In India, age at marriage is comparatively low and the high level of fertility is being attributed to the low age at marriage. Yet in recent years age at marriage has shown a rising trend and Goyal (1975) has demonstrated that for the country as a whole mean age at marriage for females has increased from 16.1. to 17.2 years and for males from 21.4. to 22.2 years during the decade 1961-71. He further observes that this increase is evident in both rural and urban areas. The increase has been of a slightly higher order in urban than rural areas. With a view to raise the age at marriage, the Government of India in its National Population Policy statement has stated that the minimum age at marriage should be raised to 18 years for girls and 21 years for boys. It further stated that "Raising the age of marriage will not only have a demonstrable demographic impact but will also lead to more responsible parenthood and help to safeguard the health of the mother and the child" (Mitra 1978).

Therefore, the main objective of the present chapter was to examine the actual age at marriage in rural communities and to identify some of the socio-economic factors that were associated with the age at marriage. While analysing these varibales an effort will be made to examine the relationship between age at marriage and fertility performance of the respondents. Lastly, an attempt would also be made to understand the intra-and inter-village comparison, if any in the age at marriage.

Since marriage is an important demographic variable accounting for the variations in fertility, it becomes important to understand the ages at which marriages take place in rural areas. Table 3.1 gives the distribution of respondents by their age at marriage.

Table 3.1 indicates that nearly one fifth of the women in these villages were married at a very young age, i.e., at 14 or less than 14, and almost one third of them were married between the ages of 15 to 17. Another one third of them were married between the ages of 18 to 20 years, whereas, the percentage of those who were married above the age of 20 years was only 14. The average age at marriage for the total sample was 17.80 years, whereas, it was 17.70, 17.88 and 17.82 respectively, for villages Ugala, Ghagga and Ghudani Kalan.

An inter-village comparison does not reveal any significant variations although the proportion of women who were married at age 14 or at less than 14 years was higher at Ugala and much lower at Ghagga. Similarly, the proportion of women who got married between the ages of 15 to 20, was higher for Ghagga but lower for Ghudani Kalan.

The data, on the whole, indicate that the villages in Punjab and Haryana have more or less the same pattern of marriage and there is no significant variation in age at marriage in these villages.

Mean age at marriage for the rural areas of Punjab was reported (Mitra 1978) to be 17.41 years in 1961 and 18.3 years in 1971. But, in the present study, which was conducted in 1980-81, mean age at marriage has been found to be slightly lower in all the three villages than what it was observed to be

TABLE 3.1

Distribution of respondents by their age at marriage

Age at marriage (in years)	Ugala		Ghagga		Ghudanl Kalan		Grand Total	
	No. of respondents	Percentage	No. of respondents	Percentage	No. of respondents	Percentage	No. of respondents	Percentage
14 & below	59	24.18	51	18.35	69	22.10	179	21.49
15—17	72	29.51	96	35.53	103	33.12	271	32.53
18—20	81	33.20	92	33·09	91	29.26	264	31.69
21—23	21	8.61	30	10.79	29	9.32	80	9.60
24 and above	11	4.50	9	3.24	19	6.11	39	4.68
Total	244	100.00	278	100.00	311	100.00	833	99.99
Mean age at marriage	17.70		17.88		17.82		17.80	

in 1971 for the rural areas as a whole. It is true that the age at marriage has been showing a rising trend during the past 50 years but it appears that the real rise in the age at marriage has taken place during the past 30 years, and it is generally the younger women below the age of 40 years who have comparatively higher age at marriage. It is, therefore, useful to analyse age at marriage by the various cohorts of women. But in the absence of such data, an attempt has been made to analyse age at marriage by the present age of the women. Therefore, Table 3.2 gives the distribution of respondents by their present age and their age when they were married.

In Table 3.2, the highest mean age at marriage, i.e., 18.57 years was noticed for those women who were young and less than 30 years of age while mean age at marriage showed consistent lowering down with an increase in the present age of the women and it was 17.09 for those respondents who were above 40 years of age.

Similar trend was observed in each village and the difference in the mean age at marriage between the younger and older women ranged from one to two years.

An inter-village comparison reveals some variations in the mean age at marriage when given the same present age of the women although the variations were not very wide. For example, for those women who were below 30 years of age, mean age at marriage was higher for Ghudani Kalan and a little lower for Ghagga while Ugala falls at the middle level. Similar variations were observed for the next age group, i.e., 30-40 years, whereas, for those women who were above 40 years of age, age at marriage was higher for Ghagga and lower for Ghudani Kalan.

Thus, the mean age at marriage for the women in the persent sample was slightly lower than the average reported for the rural areas of the state as a whole and this may be because nearly fifty per cent of the women in the present study were above the age of 40 and such women were married at a comparatively younger age.

Age at Marriage and Education

One of the important variables effecting age at marriage is the

TABLE 3.2

Distribution of respondents by their present age and mean age at marriage.

Present age of the respondents (in years)	Ugala		Ghagga		Ghudani Kalan		Grand Total	
	Mean-age at marriage	No. & % of respondents	Mean-age at marriage	No. & % of respondents	Mean-age at marriage	No. & % of respondents	Mean-age at marriage	No. & % of respondents
15—29	18.50	72 (29.51)	18.32	84 (30.22)	18.92	73 (23.47)	18.57	229 (27.49)
30—39	17.75	65 (26.64)	18.39	79 (28.42)	18.50	78 (25.09)	18.24	222 (26.65)
40 and above	17.11	107 (43.85)	17.20	115 (41.36)	16.99	160 (51.45)	17.09	382 (45.86)
Total	17.70	244 (100.)	17.88	278 (100.)	17.82	311 (100.)	17.80	833 (100.)

(Percentages are given in Parentheses.)

educational level of the population. It is contended that the higher the level of literacy of the population the higher is the age at marriage. This is specially true of females, and more so, when they are educated beyond high school level. It is so not only because all those who go in for higher education get married after the completion of their education but also because a value system that favours the production of educated women also implies assignment of roles that extend beyond that of childbearing. A clear evidence of this fact is present in the Mysore Study (1961) where it was found that the women who were educated upto middle class or above, were married on an average one year later than the females who were either illiterate or educated upto primary level. Therefore, Table 3.3 gives the distribution of respondents by their mean age at marriage and their educational attainments in these rural areas

The data in Table 3.3 show that the mean age at marriage was positively related to the educational level of the respondents. Illiterate respondents had lower age at marriage and as we move up the educational level mean age at marriage indicates rising trend and it was the highest for those respondents who were educated upto graduation or above that level for the total population as well as in each village. It is interesting to note that there was a difference of 5-7 years in the mean age at marriage between illiterate and highly educated respondents.

A comparative analysis of the three villages does not reveal any significant variation in the mean age at marriage when given the same level of educational attainments of the respondents although some minor differences appear here and there. The noticeable fact in all the three villages as well as in the total population was that mean age at marriage was slightly lower for literate women as against those who were illiterate. Since mere literacy is just not the same thing as education, therefore, these differences between the illiterates and the literates may not be of much significance.

The literacy level in the rural areas of our country is very low and it is still lower for rural females and the same trend is reflected in the present study as well. The percentage of female literacy in the rural areas of Punjab and Haryana has been

TABLE 3.3

Mean age at marriage by respondents educational level

Educational level of the respondents	Ugala		Ghagga		Ghudani Kalan		Grand Total	
	Mean-age at marriage	No. & % of respondents	Mean-age at marriage	No. & % of respondents	Mean-age at marriage	No. & % of respondents	Mean-age at marriage	No. & % of respondents
Illiterate	17.00	184 (75.41)	17.94	244 (87.77)	17.04	229 (73.63)	17.36	657 (78.87)
Literate	16.85	17 (6.97)	16.88	8 (2.88)	16.80	10 (3.21)	16.84	35 (4.20)
Primary	18.93	21 (8.61)	18.00	12 (4.32)	18.28	37 (11.90)	18.43	70 (8.40)
Middle	19.13	8 (3.28)	21.75	4 (1.44)	19.23	11 (3.54)	19.63	23 (2.76)
Secondary	20.75	12 (4.92)	21.21	7 (2.52)	21.60	20 (6.43)	21.52	39 (4.68)
Graduation	21.50	2 (0.81)	21.50	3 (1.07)	25.50	4 (1.29)	24.50	9 (1.08)
Total	17.70	244 (100.)	17.88	278 (100.)	17.82	311 (100.)	17.80	833 (100.)

(Percentages are given in parentheses.)

reported to be (census: 1981) 28.35 and 15.34, respectively. Although in the present study 21 per cent of the sample females were literate but Ghagga had much lower percentage of literate females as it was only 12.23 per cent and the proportion of women who had gone in for higher education was quite small. Nevertheless, given the number of educated women, variations in the mean age at marriage were quite significant.

It may be pointed out here that the level of literacy among respondents' husbands was much higher than the respondents themselves as more than 50 per cent of the husbands were literate or educated upto various levels, therefore, age at marriage was expected to be influenced by husband's educational level.

Womens' age at marriage is also influenced by their husbands' educational level because like women, higher education among men also means higher age at marriage and men of higher ages seek spouses in the vicinity of their own age groups. Moreover, higher educated men have a tendency to seek educated spouses and this in turn, may also influence the age at marriage. Thus Table 3.4 indicate respondents' mean age at marriage by their husbands' educational level.

It is clear from Table 3.4 that mean age at marriage for women was positively related to their husbands' educational level as it was low, i.e., 17.02 years for illiterate husbands and high, i.e., 22.09 years when husbands were educated upto graduation level. Although within each village mean age at marriage was low for illiterate husbands and high for educated husbands, still various educational categories did not reveal a consistent trend in the mean age at marriage in all these villages.

A comparative analysis of the three villages also indicates variations in the mean age at marriage when given the same level of the educational attainments of the husbands. For example, among illiterate husbands, mean age at marriage for women was higher for Ghagga and lower for Ugala, whereas, Ghudani Kalan falls in between. Similar variations were observed for other educational categories.

Age at marriage is also influenced by women's gainful

TABLE 3.4

Mean age at marriage and husband's educational level

Educational level of the husband	Ugala		Ghagga		Ghudani Kalan		Grand Total	
	Mean age at marri-age	No. & % of respon-dents	Mean age at marri-age	No. & % of respon-dents	Mean age at marri-age	No. & % of respon-dents	Mean age at marri-age	No. & % of respon-dents
Illiterate	16.58	108 (44.26)	17.44	144 (51.80)	16.92	143 (45.98)	17.00	395 (47.42)
Literate	17.90	15 (6.15)	17.53	29 (10.43)	16.62	26 (8.36)	17.27	70 (8.40)
Primary	16.40	30 (12.30)	18.63	31 (11.15)	18.60	20 (6.43)	17.80	81 (9.72)
Middle	17.65	34 (13.93)	18.14	22 (7.91)	18.92	26 (8.36)	18.18	82 (9.84)
Secondary	20.15	46 (18.85)	18.25	36 (12.95)	19.62	47 (15.11)	19.43	129 (15.49)
Graduate	18.30	5 (2.05)	22.50	5 (1.80)	23.50	12 (3.86)	22.09	22 (2.64)
Not known	15.00	6 (2.46)	17.45	11 (3.96)	16.34	34 (11.90)	16.42	54 (6.48)
Total	17.70	244 (100.)	17.88	278 (100.)	17.82	311 (100.)	17.80	833 (100.)

(Percentages are given in parentheses.)

employment, especially, when women work in occupations which require more number of years of schooling and training because such women tend to marry later than those who are not engaged in any gainful activity. In the present study the number of women engaged in gainful activities outside the home was rather small, being only 4 in Ugala, 32 in Ghagga and 9 in Ghudani Kalan. Among these women also majority of them were employed in low status occupations such as maid-servants or as agricultural labour.

Since such rural occupations do not require any formal training and most of them were illiterate, their age at marriage was just the same as that of the women who were not engaged in any activity outside the home.

Age at Marriage and Occupational Status of the Husband

Age at marriage is also influenced by the husband's occupational status. It is generally contended that those men who are engaged in higher prestige occupations which require longer periods of professional training tend to marry little later than those who are engaged in occupations of lower prestige. Since they themselves marry at a higher age, they tend to seek spouses in the vicinity of their own ages. Table 3.5 gives the distribution of respondents by their mean age at marriage and their husband's occupational* status.

In Table 3.5, occupational categories one to three include such occupations as lawyers, engineers, big farmers or businessmen and government officers, have been combined into one category because the number of respondents in each occupational category was quite small. On the whole, the data indicate positive association between husbands' occupational status and mean age at marriage of the respondents as with the exception of category one to three in Ugala all other categories' mean age

* It may be mentioned here that all the occupations were divided into seven prestige categories and the ranking of occupations into these categories was done by the villagers themselves. Therefore, in Table 3.5 the occupations followed by the husbands of the respondents were indicated by categories and the lower numbers represented higher prestige occupation and reverse was true for higher numbers.

TABLE 3.5

Mean age at marriage and husband's occupational status

Occupational prestige categories	Ugala		Ghagga		Ghudani Kalan		Grand Total	
	Mean age at marri-age	No. & % of respon-dents	Mean age at marri-age	No. & % of respon-dents	Mean age at marri-age	No. & % of respon-dents	Mean age at marri-age	No. & % of respon-dents
I-III	17.02	23 (9.43)	18.62	17 (6.12)	19.07	7 (2.25)	17.90	47 (4.64)
IV	18.91	51 (20.90)	17.80	44 (15.83)	19.03	45 (14.47)	18.60	140 (16.81)
V	18.43	73 (29.92)	17.90	98 (35.25)	18.26	124 (39.87)	18.18	245 (35.41)
VI	16.08	18 (7.38)	18.46	23 (8.27)	17.54	23 (7.40)	17.46	64 (7.68)
VII	16.45	67 (27.45)	17.85	78 (28.06)	16.85	43 (13.83)	17.12	188 (22.57)
No occupation	15.50	6 (2.46)	16.50	10 (3.60)	17.31	37 (11.90)	16.95	53 (6.36)
Not applicable	15.50	6 (2.46)	16.87	8 (2.88)	16.21	32 (10.29)	16.23	46 (5.52)
Total	17.70	244 (100.)	17.88	278 (100.)	17.82	311 (100.)	17.80	833 (100.)

at marriage for female indicated declining trend with the lowering of occupational prestige.

But the relationship between these two variables does not reveal a very clear trend within each village. For example, in village Ugala mean age at marriage was neither the highest among those respondents whose husbands were engaged in occupations of higher prestige representing from category one to three not was it the lowest among those whose husbands were engaged in occupations of lower prestige. A similar trend was noticed in Ghagga although here mean age at marriage was the highest among occupations of higher prestige but not the lowest among those respondents whose husbands were employed in occupations of lower prestige The other occupational categories also did not indicate any clear trend.

But in Ghudani Kalan this relationship was positive and mean age at marriage was the highest for those respondents whose husbands were employed in higher prestige occupations and the lowest for those respondents whose husbands were employed in low prestige occupations.

But an inter-village comparison reveals marked variations in the mean age at marriage when given the same occupational category of the husbands indicating no inter-village association between mean age at marriage and occupational status of the husband. On the whole, an inter-village and intra-village relationship between these two variables brings us to the conclusion that mean age at marriage is not much influenced by the occupational status of the husband and this may be because most of these occupations are rural in nature and are not associated with any formal or professional training.

Age at Marriage and Economic Status

The economic status of the family is another important factor that influences age at marriage. Although information on age at marriage by income of wife's or husband's family or even level of consumption is difficult to come by, Das Gupta and others (April-September 1952) in couple fertility based on the data of NSS, adopted the expenditure approach as an indicator of the level of living in view of the difficulty to get at the correct income. In the present study in order to gauge the economic status of the respondents, level of living index was prepared and an attempt was made to relate the mean age at marriage to the level of living which is a combination of various factors rather than to the simple income categories. Table 3.6 gives the distribution of respondents by their mean age at marriage and their level of living.

The table clearly reveals positive association between mean age at marriage and level of living of the respondents in the total as well as in each village as the mean age at marriage was low for those respondents who had low level of living and high for those who had higher level of living in the total as well as in each village.

A comparative analysis of data for all the three villages reveals that given the same level of living, there are variations in the mean age at marriage among the three villages although these variations are not well marked.

Age at Marriage and Caste Status

Caste is an important institution in our society as it is woven into the entire social fabric and age at marriage is also influenced by it. The differences in the mean age at marriage by caste have been noticed by some studies. For example, Agarwala (1966) reported that, on the whole, depressed castes have the lowest mean age at marriage followed by Brahmins, the warrior castes and the trading castes except in the extreme South where the Brahmin females have the lowest age at marriage. In the present study an attempt has been made to see whether mean age at marriage of the respondents has any association with their caste status. Table 3.7 shows the distribution of respondents by their mean age at marriage and their caste status.

It may be mentioned here that all the castes included in the study have been divided into four categories. Category one includes higher castes like Brahmins and Khatris, category two consists of middle castes like Jats, Aroras and Sainis etc., category three represents all the backward castes and in category four are included the scheduled castes. Therefore, in Table 3.7 the various castes are referred to as upper castes, middle or intermediate and lower castes and scheduled castes.

TABLE 3.6

Mean age at marriage and level of living

Level of living	Ugala		Ghagga		Ghudani Kalan		Grand Total	
	Mean age at marri-age	No. & % of respon-dents	Mean age at marri-age	No. & % of respon-dents	Mean age at marri-age	No. & % of respon-dents	Mean age at marri-age	No. & % of respon-dents
Low level	16.95	93 (33.11)	17.78	82 (29.50)	16.66	56 (18.01)	17.17	231 (27.73)
Medium level	18.00	56 (22.95)	17.73	93 (33.45)	17.31	63 (20.26)	17.68	212 (25.45)
High level	18.27	95 (38.94)	18.10	103 (37.05)	18.33	192 (61.73)	18.25	390 (46.82)
Total	17.70	244 (100.)	17.88	278 (100.)	17.82	311 (100.)	17.80	833 (100.)

TABLE 3.7

Mean age at marriage and caste status

Caste status	Ugala		Ghagga		Ghudani Kalan		Grand Total	
	Mean age at marri-age	No. & % of res-pondents	Mean age at marri-age	No. & % of res-pondents	Mean age at marri-age	No. & % of res-pondents	Mean age at marri-age	No. & % of res-pondents
Upper caste	18.45	103 (42.21)	18.12	76 (27.34)	18.46	26 (8.36)	18.33	205 (24.61)
Intermediate caste	17.00	12 (4.92)	18.38	59 (21.22)	18.13	196 (63.02)	18.13	267 (32.05)
Lower caste	17.83	63 (25.82)	17.73	34 (12.23)	17.48	40 (12.86)	17.70	137 (16.45)
Scheduled caste	16.54	66 (27.05)	17.44	109 (39.21)	16.56	49 (15.76)	16.98	224 (26.89)
Total	17.70	244 (100.)	17.88	278 (100.)	17.82	311 (100.)	17.80	833 (100.)

The data in Table 3.7 clearly reveals positive association between mean age at marriage and the caste status of the respondents in the total as well as in Ghudani Kalan. In village Ugala and Ghagga, although mean age at marriage was higher for respondents of upper castes and lower for those belonging to scheduled caste, there were some discrepancies between the intermediate and the lower castes in these two villages. But, on the whole, the trend clearly indicates positive association between the two.

An inter-village comparison does not reveal any significant variation in the mean age at marriage when given the same caste background of respondents.

Age at Marriage and Religion

Differences in the mean age at marriage among different religious groups are also well marked in India. But since 1931 census data by religion was discontinued, therefore, no global picture is available. But there are some surveys which indicate these differentials. For example Agarwala (1966) reported that between 1891-1931 census, Christians retained a distinct level in age at marriage for both males and females followed by Sikhs, Muslims and Jains and Hindus had the lowest age at marriage. Agarwala's findings are also corroborated by those of the Mysore Population Study. The order of the highest average age at marriage for women prevailing among Christians followed by Sikhs, Muslims and Hindus, was also reflected in the rural data of NSS conducted around this period (Das Gupta 1951) in couple fertility. Table 3.8 gives the distribution of respondents by their mean age at marriage and their religion.

In the present sample there were two major religious groups, i.e., Hindus and Sikhs and in Ghagga aad Ghudani Kalan there were a few Muslim families, therefore, differences between the Hindus and the Sikhs only are noteworthy. The data indicate that the Hindus had lower mean age at marriage than the Sikhs in the total sample as well as in each village, although the differences were rather small. Further, given the small number of Muslims in the two villages, their mean age at marriage was comparatively low and less than the other two religious groups.

TABLE 3.8

Mean age at marriage and religion

Religion	Ugala		Ghagga		Ghudani Kalan		Grand Total	
	Mean age at marri-age	No. & % of res-pondents	Mean age at marri-age	No. & % of res-pondents	Mean age at marri-age	No. & % of res-pondents	Mean age at marri-age	No. & % of res-pondents
Hindu	17.51	196 (80.33)	17.81	182 (65.47)	17.83	43 (13.83)	17.67	421 (50.54)
Muslim	—	—	16.50	4 (1.44)	15.83	9 (2.89)	16.08	13 (1.56)
Sikh	18.50	48 (19.67)	18.07	92 (33.09)	17.89	259 (83.28)	18.00	399 (47.90)
Total	17.70	244 (100.)	17.88	278 (100.)	17.92	311 (100.)	17.80	833 (100.)

An inter-village comparison also did not reveal any significant variations in the mean age at marriage when given the same religious background of the respondents.

Lastly, an attempt was made to see whether there is any association between the mean age at marriage and the modernity level of the respondents. As already stated, for measuring the level of modernity of the respondents, an index of tradition-modernity was prepared and the respondents were divided into three categories namely, modern, moderately modern and traditional. Table 3.9 gives the distribution of respondents by their mean age at marriage and level of modernity.

In Table 3.9 the highest mean age at marriage was observed among modern respondents, whereas, the lowest for traditional respondents in the total as well as in each village. Further, an important point to note in terms of inter-village comparison was that among modern respondents mean age at marriage was higher in Ghudani Kalan and lower in Ugala and Ghagga stands in the middle. On the other hand, among traditional respondents mean age at marriage was low for Ghudani Kalan and higher for Ghagga indicating difference between the modern and traditional respondents to be 2.26 years for Ghudani Kalan followed by 1.48 years in Ugala and 0.31 year for Ghagga.

Having discussed the variations in the age at marriage with socio-economic characteristics of the respondents, it is now proposed to analyse mean fertility of women by their age at marriage.

Age at Marriage and Fertility

Age at marriage is considered to have an important bearing on the fertility behaviour. The general hypothesis is that the higher the age at marriage the greater will be the influence in depressing fertility. Age at marriage determines the duration of marriage, and through duration, it influences fertility. In addition to this, age at marriage also affects fertility through higher education and gainful employment by exposing the women to the wider world and by developing the sense of responsible parenthood in them.

Various studies in India have indicated a negative association between age at marriage and fertility. They show that average

TABLE 3.9

Mean age at marriage and level of modernity

Level of modernity	Ugala		Ghagga		Ghudani Kalan		Grand Total	
	Mean age at marriage	No. & % of respondents	Mean age at marriage	No. & % of respondents	Mean age at marriage	No. & % of respondents	Mean age at marriage	No. & % of respondents
Traditional	17.30	123 (50.41)	17.92	93 (33.45)	16.69	109 (35.05)	17.27	325 (39.02)
Moderately modern	17.88	91 (37.30)	17.67	121 (43.53)	18.15	131 (42.12)	17.91	343 (41.18)
Modern	18.78	29 (11.89)	18.23	64 (23.02)	18.95	71 (22.83)	18.64	164 (19.68)
Not recorded	19.50	1 (0.40)	—	—	—	—	19.50	1 (0.12)
Total	17.70	244 (100.)	17.88	278 (100.)	17.82	311 (100.)	17.80	833 (100.)

number of children born by age at marriage decrease, with discrete increases in marriage age. For instance, the Mysore Study (1961) shows that, when women aged 45-54 years with unbroken marriage were divided into four groups according to their age at marriage as under 14, 14-17, 18-21 and 22 years and above, the number of children born to these cohorts in urban areas were 6.7, 6.0, 5.7 and 4.6. Therefore, Table 3.10 gives the distribution of respondents by their age at marriage and mean number of live-births.

Table 3.10 clearly indicates close association between age at marriage and mean number of live-births in every village as well as for the total sample. Mean number of live-births was the highest among women who were married before the age of 14 years and the number shows consistent decline with an increase in the age at marriage and there was a difference of nearly two children between the lower and higher age at marriage. The data further indicate that given the same age at marriage, inter-village differences in the mean number of live-births were not very significant execpt in the last category of 24 years and above where mean number of live-births was comparatively much higher in Ugala. But the number of such cases was rather small for making any generalizations.

Some scholars (Glass and Grebenick 1954) have suggested that age at marriage is a less obvious determinant of family size. But Agarwala (1962) considers it an important independent determinant of fertility and says that in countries like India where measures for family limitation are not in general use, the age at which marriage is contracted can have significant effect on the completed size of the family. It thus becomes necessary to examine the influence of age at marriage on fertility by controlling other variables. Therefore, Table 3.11 gives the distribution of respondents by their age at marriage, duration of marriage and mean number of live-births.

It may be mentioned here that the performance of marriage is not necessarily followed by cohabitation of the couple. It is generally followed by another ceremony called 'Gauna' or 'Vida' and it is only after this ceremony that the newly married couple begins living together. It is also called the age of effective marriage and it is this age of effective marriage which

TABLE 3.10

Mean age at marriage and mean number of live-births

Age at Marriage (in years)	Ugala		Ghagga		Ghudani Kalan		Grand Total	
	Mean No. of live-births	No. & % of respondents	Mean No. of live-births	No. & % of respondents	Mean No. of live-births	No. & % of respondents	Mean No. of live-births	No. & % of respondents
14 & below	6.12	58 (24.68)	5.13	50 (18.32)	5.47	66 (22.00)	5.58	174 (21.53)
15–17	4.95	68 (28.94)	5.07	95 (34.80)	4.50	100 (33.33)	4.82	263 (32.55)
18—20	4.20	81 (34.47)	4.65	91 (33.33)	3.23	89 (29.97)	4.02	261 (32.30)
21—23	2.95	20 (8.51)	3.48	28 (10.26)	3.0)	27 (9.00)	3.17	75 (9.28)
24 and above	4.40	8 (3.40)	3.33	9 (39.2)	2.68	18 (6.00)	3.24	35 (4.34)
Total	4.02	235 (100.)	4.72	273 (100.)	4.12	300 (100.)	4.53	808 (100.)

TABLE 3.11

Mean age at marriage, duration of marriage and mean number of live-births

Age at marriage (in yrs.)	Ugala 0-19 Yrs.		Ghagga 0-19 Yrs.		Ghudani Kalan 0-19 Yrs.		Ugala 20+Yrs.	
	Mean No. of live-births	No. & % of respondents	Mean No. of live-births	No. & % of respondents	Mean No. of live-births	No. & % of respondents	Mean No. of live-births	No. & % of respondents
14 yrs. & below	4.11	15 (12.93)	2.70	20 (13.70)	3.64	11 (7.86)	7.10	43 (36.13)
15—17	3.48	31 (26.72)	2.89	48 (32.88)	3.31	45 (32.14)	6.38	37 (31.09)
18—20	2.70	48 (41.38)	2.81	56 (38.36)	2.41	53 (37.86)	6.10	33 (27.73)
21—23	2.41	17 (14.66)	2.04	19 (13.01)	2.61	18 (12.86)	5.35	3 (2.52)
24 and above	2.80	5 (4.31)	2.33	3 (2.05)	1.61	13 (9.28)	5.35	3 (2.52)
Total	3.09	116 (100.)	2.70	146 (100.)	2.75	140 (100.)	6.52	119 (99.99)

Age at marriage (in yrs.)	Ghagga 20+Yrs.		Ghudani Kalan 20+Yrs.		Total 0-19 Yrs.		Total 20+Yrs.	
	Mean No. of live-births	No. & % of respondents	Mean No. of live-births	No. & % of respondents	Mean No. of live-births	No. & % of respondents	Mean No. of live-births	No. & % of respondents
14 Yrs. & below	8.50	30 (23.62)	6.60	55 (34.37)	3.39	46 (11.44)	6.96	128 (31.53)
15—17	6.15	37 (37.01)	5.58	55 (34.37)	3.21	124 (30.85)	6.00	139 (34.24)
18—20	7.75	35 (27.56)	4.45	36 (22.50)	2.64	157 (39.05)	6.09	104 (25.62)
21—23	6.50	9 (7.09)	4.45	9 (5.63)	2.35	54 (13.43)	5.46	21 (5.17)
24 and above	3.50	6 (4.72)	3.00	5 (3.13)	2.00	21 (5.22)	3.72	14 (3.44)
Total	7.05	127 (100.)	5.33	160 (100.)	2.83	402 (100.)	6.21	406 (100.)

has an effect on fertility. For the present study information on the date of consummation of marriage was not available for all the respondents and to avoid this difficulty, we calculated the duration of marriage from the age of consummation of marriage where such information was available and in other cases from the 14th year when marriage took place prior to that age.

It is clear from Table 3.11 that for every duration category as well as for the total, the mean number of live-births was higher for those respondents who were married either below 14 years of age or between 15-17 years of age, and was lower for those who were married above the age of 20, although within each duration category there were some inconsistencies. This relationship is more significant in the duration category of 20 years and above where majority of the women had either completed their reproductive span or are near completion in all the villages as well as in the total indicating thereby that when controlled for duration, age at marriage is an important determinant of fertility.

Summing up, it can be observed that although mean age at marriage for the total population was 17-18 years, still younger women below the age of 30 were married at a comparatively higher age in all these villages indicating a rising trend in the age at marriage among younger women. Age at marriage was positively associated with the socio-economic level of the respondents and negatively associated with the fertility level of the respondents.

REFERECNES

1. Agarwala, S.N. 1974: India's Population Problem, New Delhi, Tata McGraw Hill Publishing Company.
2. Agarwala, S.N. 1962: Age at Marriage in India, Bombay, Kitab Mahal.
3. Census Commissioner 1981: Census of India, Series 1, Paper No. 3 of 1981.
4. Das Gupta & Others 1952: Couple Fertility, NSS Report, 7.
5. Davis, K. 1964: "Population Policy: Will Current Programmes Succeed?", Population Evaluation and Birth Control, Garrett Harden, Freeman and Company, San Francisco.

6. Glass, D.V. & Grebenick 1954: "The Trend and Patterns of Fertility in Great Britain", Papers of the Royal Commission on Population, London, HMSO, 2 Vols.
7. Goyal, R.P. 1975: Shifts in Age at Marriage in India between 1961-71, Demography India, Vol. IV, 2.
8. Jain, S.P. 1975: A Status Study on Population Research in India, Vol. II, New Delhi, Tata McGraw Hill Publishing Company.
9. Mitra, Asok 1978: India's Population—Aspects of Quality and Control, Vol. I, New Delhi, Abhinav Publications.
10. Reddy, P.H. & Gopal, Y.S. 1980: Socio-Economic Factors and Age at Marriage, Journal of Family Welfare, Vol. xxv, 3.
11. Singh, K.P. 1979: Status of Women and Publication Growth in India, New Delhi, Munsiram Manohar Lal Publishers.
12. United Nations 1961: Mysore Population Study, New York,

4

DEMOGRAPHIC FACTORS AND FERTILITY

In the present chapter an attempt has been made to examine some of the demographic variables that are related to fertility behaviour. Under demographic variables are included the present age of the respondents, their age at the birth of the first child and total duration of married life. Although it is a study of three different villages chosen with a specific purpose, the data have been analysed separately for each village and conclusions are also drawn for the village community as a whole as there are no siginficant variations in the basic characteristics of these villages.

As already mentioned, the basic purpose of the study was to examine the various correlates of fertility, therefore, childless women have been excluded from the analysis as in their case there was nothing to explain in terms of socio-economic determinants of fertility. The proportion of widows, divorced and separated women was 8, 7 and 12 per cent in Ugala, Ghagga and Ghudani Kalan respectively and these women have been included in the study.

Table 4.1 gives the distribution of respondents according to their present level of fertility in three villages as well as in each village, separately.

The data in this table indicate that the mean number of pregnancies was 4.54 for the total sample, but in terms of each village community, it was 4.80 for Ugala, 4.75 for Ghagga and 4.09 for Ghudani Kalan. The data further indicate that the mean number of live-births was also in the same order being the highest in Ugala and the lowest in Ghudani Kalan and Ghagga being midway between these two figures.

Another noticeable fact was that the number of male births

TABLE 4.1

Distribution of respondents by their mean number of pregnancies, live-births, deaths and pregnancy wastage

Fertility level	Ugala	Ghagga	Ghudani Kalan	Total
Mean No. of pregnancies	4.80	4.75	4.09	4.54
Mean No. of live-births	4.82	4.72	4.12	4.53
Mean No. of male births	2.50	2.47	2.31	2.42
Mean No. of female births	2.32	2.25	1.81	2.11
Mean No. of deaths	0.68	0.72	0.83	0.75
Average pregnancy wastage	0.16	0.09	0.11	0.10
Average age of the respondents	38.12	37.82	38.66	38.22
Average duration of married life	20.47	19.48	20.16	20.02
Total No. of respondents	231	257	291	779

was higher than the female births in all the three villages and the difference was more marked in Ghudani Kalan as compared to the other two villages. In terms of percentage, there were 51.90 and 48.10 per cent male-female births in Ugala, 52.36 and 47.64 per cent for Ghagga and 56.09 and 43.91 per cent male-female births for Ghudani Kalan.

The data further reveal that although mean number of deaths was 0.75 for the total sample but in terms of inter-village comparison the trend was opposite to that of live-births as mean number of deaths was the highest in Ghudani Kalan, followed by Ghagga and it was the lowest in Ugala. It was also observed that in Ugala female deaths were more than male deaths but in the other two villages reverse was the case. In terms of percentage in each village, of the total number of live-births 14.10 per cent, 15.70 and 20.12 per cent, respectively, had died in Ugala, Ghagga and Ghudani Kalan. Pregnancy wastage was not very high in these villages although it was comparatively higher four Ugala, followed by Ghudani Kalan and lowest for Ghagga. Again, in terms of percentage it was three, two and two per cent, respectively, for Ugala, Ghagga and Ghudani Kalan. Khanna Study (1971) reported pregnancy wastage of 13.6 per cent which also included abortions and still-births but there are not many studies at all-India level to make a comparison. But in comparison to Khanna Study, the low pregnancy wastage may be because of under reporting on the part of women who generally do not like to report about still-births.

Thus, on the whole, it was found that the number of children born ranged from 0 to 11 in these villages. In Ugala, 46 per cent of the women had 1-4 live-births and the remaining 54 per cent had five or more live-births. In Ghagga, 48 per cent of them had 1-4 live-births, whereas in Ghudani Kalan 56 per cent of the women had 1-4 live-births, and the remaining 44 per cent had five or more than five live-births. Thus, in terms of parity, more women in Ugala as compared to other two villages had higher parity births.

Taking into account the present average age of the respondents which was 38.12 years, 37.82 and 38.66 years respectively, for Ugala, Ghagga and Ghudani Kalan and their marriage duration (20.47, 19.48, 20.16 years, respectively, for all the three villages), it may be stated that the present level of fertility was comparatively higher for Ugala and lower for Ghudani Kalan, Ghagga coming in between the two.

The number of live-births is closely related to the present age of the women and it is expected that younger women would have lesser number of children as compared with women in advance ages as they have not yet completed their reproductive span. Therefore, Table 4.2 gives the distribution of respondents by their present age and the mean number of live-births.

Table 4.2 clearly indicates that the mean number of live-births was the lowest among younger women. The number of

TABLE 4.2

Mean number of live-births by respondents present age

Present age of the respondents (in years)	Ugala		Ghagga		Ghudani Kalan		Total	
	Mean No. of live-births	No. & % of respondents	Mean No. of live-births	No. & % of respondents	Mean No. of live-births	No. & % of respondents	Mean No. of live-births	No. & % of respondents
15-19	1.00	4 (1.70)	0.36	14 (5.13)	—	3 (1.00)	0.43	21 (2.60)
20-24	1.55	31 (13.19)	1.60	34 (12.45)	1.22	27 (9.00)	1.47	92 (11.39)
25-29	2.90	33 (14.04)	2.50	39 (14.29)	2.37	43 (14.33)	2.57	115 (14.23)
30-34	4.50	45 (19.15)	4.09	54 (19.78)	3.95	45 (15.00)	4.17	144 (17.82)
35-39	5.47	19 (8.09)	5.50	23 (8.42)	4.41	32 (10.67)	5.02	74 (9.16)
40-44	5.57	26 (11.06)	6.00	27 (9.89)	4.60	41 (13.67)	5.27	94 (11.63)
45 and above	6.95	77 (32.77)	7.50	82 (30.04)	5.50	109 (36.33)	6.53	268 (33.17)
Total	4.82	235 (100.)	4.72	273 (100.)	4.12	300 (100.)	4.53	808 (100.)

(Percentages are given in parentheses.)

live-births increases consistently as the age advances and it was the highest among women who were 45 years or above, in all the three villages as well as in each village separately.

An inter-village comparison indicates that given the same present age of the respondents mean number of live-births was higher upto the age of 35 years for Ugala followed by Ghagga and the lowest for Ghudani Kalan, but above the age of 35 years including respondents who were of completed fertility mean number of live-births was higher for Ghagga and somewhat lower for Ugala and the lowest for Ghudani Kalan. Thus, on the whole the data shows that given the same present age of the respondents fertility level was again higher in Ugala and Ghagga but comparatively lower for Ghudani Kalan

Age at the birth of the first child is another important variable that influences fertility performance. It is maintained that women are more fertile in their younger ages, and those who get married early and start producing children at a younger age are likely to produce more children than those who start producing children at a later age.

Table 4.3 gives mean number of live-births by respondent's age at first birth.

The table shows that those respondents who had started producing children at a comparatively younger age, i.e., below 16, had higher proportion of live-births and the proportion of live-births declined with increasing age at the time of first birth and it was the lowest when the age at first birth was above 22 years for the total as well as for each village. Although there were some inconsistencies with increasing age, but, on the whole the data do indicate the effect of age at first birth on the mean number of live-births. An inter-village comparison revealed some variations in the mean number of live-births when given the same age at first birth though these were not very significant except in the last age category of 27 years and more, where the mean number of live-births was much lower in Ghudani Kalan as compared with the other two villages.

Length of reproductive span of a woman is the most important physiological factor determining fertility. It is generally held that as the duration of married life increases, the number of children born also increases, unless the couples practice family

TABLE 4.3

Mean number of live-births by respondent's age at first birth

Age at first birth (in years)	Ugala		Ghagga		Ghudani Kalan		Grand Total	
	Mean No. of live-births	No. & % of respondents	Mean No. of live-births	No. & % respondents	Mean No. of live-births	No. & % of respondents	Mean No. of live-births	No. & % of respondents
13-14	6.88	8 (3.46)	5.50	4 (1.56)	7.40	5 (1.72)	6.71	17 (2.18)
15-16	6.74	27 (11.69)	5.70	27 (10.51)	5.81	16 (5.50)	6.16	70 (8.99)
17-18	5.14	42 (18.18)	5.06	54 (21.01)	5.24	78 (26.80)	5.16	174 (22.34)
19-20	4.52	57 (24.68)	5.00	54 (21.01)	4.48	61 (20.96)	4.65	172 (22.08)
21-22	4.51	43 (18.62)	4.80	52 (20.23)	3.34	65 (22.34)	4.13	160 (20.54)

23-24	3.97	30 (12.99)	4.35	35 (13.62)	3.59	32 (11.00)	3.98	97 (12.45)
25-26	4.87	15 (6.49)	5.42	19 (7.39)	3.11	19 (6.53)	4.44	53 (6.80)
27+	4.11	9 (3.89)	5.17	12 (4.67)	2.33	15 (5.15)	3.72	36 (4.62)
Total	4.91	231 (100.)	5.02	257 (100.)	4.26	291 (100.)	4.70	779 (100.)

(N. varies because the number of respondents who did not have any pregnancy have been deleted.)

planning. Table 4.4 gives the distribution of respondents according to their duration of marriage and the mean number of live-births.

TABLE 4.4

Distribution of respondents by their duration of marriage and mean number of live-births

Duration of marriage	Mean number of live-births			
	Ugala	Ghagga	Ghudani Kalan	Total
0-9 years	1.76	1.52	1.54	1.59
10-19	4.28	3.92	3.83	3.99
20 years and above	6.52	7.05	5.33	6.21

It is clear from Table 4.4 that mean number of live-births was small with shorter duration of married life and it increased with an increase in the duration of married life, and it was the highest among those women who were married for more than 20 years for the total as well as for each village.

The conclusion is that although given the present average age of the respondents and their total marriage duration mean number of live-births was little higher for Ugala and lower for Ghudani Kalan, Ghagga falling in between the two; but when duration of marriage was split into various groups, women in the marriage duration of 20 years and more in Ghagga had the highest number of live-births.

REFERENCES

1. Govt. of India 1971: Fertility Differentials in India, Office of the Registrar General and Census Commissioner, New Delhi.
2. Wyon, John W. & Gordon, John E. 1971: The Khanna Study: Population Problems in the Rural Punjab, Cambridge, Harvard University Press.

5

INFANT AND CHILD MORTALITY AND FERTILITY

MORTALITY rate especially infant and child mortality, is an important determinant of fertility. It is generally argued that when mortality rate is high, fertility level also goes up as parents desire that their children should survive and when greater number of them die, they produce more to ensure the survival of some. In other words, parents do not merely want babies but babies who should survive.

It is generally stated that in developing countries high rate of infant and child mortality is one of the deterrent factors in the adoption of small family size norms because successful reproduction requires high fertility to off set high mortality rate. The U.N. Department of Economic and Social Affairs (1975) notes, "Evidence accumulates that reduction of infant mortality may be a necessary prerequisite to the acceptance of family planning. Couples will not wish to prevent pregnancies until they have some assurance that the children they already have will survive." Thus decline in mortality rate appears to be a necessary precondition for the acceptance of small family norm in developing countries.

The level of mortality in itself is greatly influenced by various other factors. There is no denying the fact that part of the variation in mortality is related to biological characteristics like age, sex and order of birth of male and female children etc., but it is also largely influenced by the socio-economic variables. It is generally maintained that the incidence of mortality is high among uneducated and low income group people, or, those people who have low level of living as against those who are educated and enjoy higher level of living. It is not because education or wealth per se has anything to do with mortality but because educated people have better knowledge of the availability and use of medical facilities. They are more concerned with hygine and sanitary conditions than the uneducated.

The present chapter is, therefore, an attempt to investigate the extent, especially, of infant and child mortality rate, in these rural areas. An attempt would also be made to examine the relationship between infant and child mortality and fertility. Lastly, some of the socio-economic correlates of infant and child mortality rate would also be examined.

General Level of Mortality

Table 5.1 gives the distribution of respondents by the extent of general mortality.

The data in Table 5.1 reveal that although of the total number of live-births, 16.47 per cent had resulted in deaths, however, this percentage was different for each village.

In Ugala, of the total number of live-births, 14.10 per cent had resulted in deaths, whereas, this percentage was 15.10 for Ghagga and 20.12 for Ghudani Kalan, indicating thereby that the general level of mortality was comparatively low in Ugala, followed by Ghagga but higher in Ghudani Kalan.

In terms of sex differentials in general mortality, the data did not reveal any uniform trend in all the villages. In the total sample as well as in Ugala, the percentage of female deaths was higher than the male deaths, but reverse was the case for Ghagga and Ghudani Kalan. It also indicated that the difference between male and female deaths was very small in Ghudani Kalan and Ugala but well marked in Ghagga. Thus, on the whole, sex differences in general mortality are there, though there is no uniform trend. Jain (1979) has also pointed out that sex differentials in mortality are weak in India. In advanced countries, males clearly show a higher mortality rate than female because mortality by sex is affected by two factors, namely, the innate biologically superior stamina of the female child and her cultural status in society. In a less developed country like India, a lower status accorded to female children and her innate biologically superior stamina, the two factors

TABLE 5.1

Extent of general mortality

	Ugala			Ghagga			Ghudani Kalan			Grand Total		
	M	F	T	M	F	T	M	F	T	M	F	T
Total No. of live-births	589	546	1135	676	615	1291	693	543	1236	1958	1704	3662
Total No. of deaths	78	82	160	106	89	195	126	122	248	310	293	603
Percentage of deaths by sex	48.75	51.25	100.	54.36	45.64	100.	50.81	49.19	100.	51.41	48.49	100.
Percentage of total deaths	13.24	15.02	14.10	15.69	14.47	15.10	18.18	22.47	20.12	15.83	17.19	16.47
Total No. of respondents			231			257			291			779

pull in opposite directions as a result weak sex differentials are observed.

Age is an important variable in the analysis of mortality, for death rates vary with age and sex. Table 5.2 gives the incidence of mortality by age and sex.

Table 5.2 shows significant variations in mortality by age. It shows that the highest proportion of deaths was that of infants below the age of one year, followed by children in the age group of 1-4 years and the lowest proportion was that of others above the age of 5 years for the three as well as for each village separately.

Although the general trend was the same in all the villages, inter-village variations were quite significant in the first two age categories. For example, in the case of infant deaths, the percentage was comparatively much higher for Ghagga and lower for Ghudani Kalan. But in case of child deaths the trend, was the reverse, being that highest in Ghudani Kalan followed by Ugala and the lowest in Ghagga. But there were no significant variations in the older age groups.

In terms of sex differentials in each age group, the trends were quite consistent with the expected tendencies as the mortality rate was higher for males as against females in the age group of 0-1 year but opposite was the trend in the age group of 1-4 years in all these villages. In the older age groups, except for Ugala, again, male mortality was higher than female mortality.

Having discussed the trends in the overall mortality in general among respondents, the remaining analysis would be confined to only infant and child mortality. It is because firstly, child mortality is an important determinant of overall mortality, and, secondly, because infant and child mortality is closely related to fertility.

Infant and Child Mortality

Since the highest proportion of deaths comprise that of infant and children, it becomes necessary to understand the extent of these deaths by sex and age differentials. Table 5.3 gives the extent of infant and children deaths by age and sex.

Table 5.3 reveals that in the total the percentage of

TABLE 5.2

Incidence of general mortality by age and sex

Age at the time of death	Ugala			Ghagga		
	No. & % of male deaths	No. & % of female deaths	Total	No. & % of male deaths	No. & % of female deaths	Total
Less than 1 year	56 (71.80)	48 (58.54)	104 (65.00)	76 (71.70)	63 (70.78)	139 (71.28)
1—4 years	17 (21.79)	24 (29.27)	41 (25.62)	20 (18.86)	22 (24.72)	42 (21.53)
5—15 years	4 (5.13)	8 (9.76)	12 (7.50)	5 (4.72)	4 (4.50)	9 (4.61)
15 and above	1 (1.28)	2 (2.24)	3 (1.87)	5 (2.56)	—	5 (2.56)
Total	78 (48.75)	82 (51.25)	160 (100.)	106 (54.36)	89 (45.64)	195 (100.)

TABLE 5.2—(*Contd.*)

Age at the time of death	Ghudani Kalan			Grand Total		
	No. & % of Male deaths	No. & % of female deaths	Total	No. & % of male deaths	No. & % of female deaths	Total
Less than 1 year	71 (56.35)	68 (55.74)	139 (56.05)	203 (65.48)	179 (61.09)	382 (63.35)
1—4 years	42 (33.33)	45 (36.89)	87 (35.08)	79 (25.49)	91 (31.06)	170 (28.19)
5—15 years	7 (5.56)	4 (3.27)	11 (4.43)	16 (5.16)	16 (5.46)	32 (5.31)
15 and above	6 (4.76)	5 (4.10)	11 (4.43)	12 (3.87)	7 (2.39)	19 (3.15)
Total	126 (50.81)	122 (49.19)	248 (100.)	310 (51.41)	293 (48.59)	603 (100.)

TABLE 5.3

Infant and child deaths by age and sex

Age at the time of death	Ugala			Ghagga		
	No. & % of male deaths	No. & % of female deaths	Total No. of deaths	No. & % of male deaths	No. & % of female deaths	Total No. of deaths
0 to 1 month	28 (38.36)	17 (23.61)	45 (31.03)	46 (47.92)	34 (40.00)	80 (44.20)
1 month to 1 year	28 (38.36)	31 (43.06)	59 (40.69)	30 (31.25)	29 (34.12)	69 (32.60)
1 year to 4 years	17 (23.29)	24 (33.33)	41 (28.28)	20 (20.83)	22 (25.88)	42 (23.20)
Total	73 (50.34)	72 (49.66)	145 (100.00)	96 (53.04)	85 (46.96)	181 (100.00)

Age at the time of death	Ghudani Kalan			Grand Total		
	No. & % of male deaths	No. & % of female deaths	Total No. of deaths	No. & % of male deaths	No. & % of female deaths	Total No. of deaths
0 to 1 month	38 (33.63)	32 (28.32)	70 (30.97)	112 (39.72)	83 (30.74)	195 (35.33)
1 month to 1 year	33 (29.20)	36 (31.86)	69 (30.53)	91 (32.27)	96 (35.56)	187 (33.88)
1 year to 4 years	42 (37.17)	45 (39.82)	87 (38.50)	79 (28.01)	91 (33.70)	170 (30.79)
Total	113 (50.00)	113 (50.00)	226 (100.00)	282 (51.09)	270 (48.91)	552 (100.00)

deaths in the neo-natal period, i.e., within the first four weeks of life after birth, was the highest and it was followed by the post neo-natal period, i.e., from the 2nd month to the completion of the first year of life and somewhat lower in the age group of 1-4 years. But no uniform trend was observed for each village separately. In fact, Ugala and Ghudani Kalan presented striking variations, whereas, Ghagga had the same trend as was observed for the total sample population. In Ugala, the highest percentage of deaths was observed in the post neo-natal period followed by neo-natal and the age group of 1-4 years, whereas, in Ghudani Kalan, the highest percentage of deaths was noted in the age group of 1-4 years. Thus, sufficiently large proportion of deaths in the post neo-natal period and in the age group of 1-4 years, indicate that more of environmental or exogenic factors rather than endogenic factors had caused these deaths.

Unlike age differentials, sex differentials in the infant and child mortality were more clear and consistent. In the neo-natal period, male deaths were higher than the female deaths in the total as well as in each village whereas, in the post neo-natal period as well as in the age group of 1-4 years, female deaths were higher than male deaths. Bhattacharjee (1979) also noted these differentials in his study at the national level. According to him, the infant mortality rate for male births is higher than the IMR for female births, and the same finding holds good for most of the states The probability of survival for females in the age group of 1-4 years is found to be lower than that for males in the same age groups. The explanation may be that though females are biologically stronger at birth and are better capable of resisting environmental hazards, they suffer a major set-back because of social neglect in the age group of 1-4 years.

Infant and Child Mortality and Fertility

It is well established by now that there is a direct relationship between fertility and mortality and the higher the fertility, the higher is the rate of infant mortality and vice-versa. Therefore, Table 5.4 gives the number of infant deaths by the number of live-births among these respondents.

Table 5.4 clearly indicates direct association between number of live-births and number of infant and child deaths in each

TABLE 5.4

Number of childern dead by the number of children ever born

Number of ever born children	Ugala				Ghagga				Ghudani Kalan				Total			
	No. of children dead			Average number of children	No. of children dead			Average number of children	No. of children dead			Average number of children	No. of children dead			Average number of children
	1-2	3-4	5+		1-2	3-4	5+		1-2	3-4	5+		1-2	3-4	5+	
1—2	7	—	—	1.00	4	—	—	1.00	2	—	—	1.00	13	—	—	1.00
3—4	10	1	—	1.45	19	2	—	1.48	22	1	—	1.26	51	4	—	1.38
5—6	20	—	1	1.38	25	3	—	1.43	41	3	—	1.64	86	6	1	1.52
7 and above	29	8	5	2.19	39	10	3	2.04	28	17	4	2.43	96	35	12	2.22
Total	66	9	6	1.78	87	15	3	1.72	93	21	4	1.88	246	45	13	1.80

village. The average number of child deaths was 1.0 when the women had one to two live-births and this average increased with an increase in the number of live-births and it was more than two among women who had seven or more live-births. It further shows that the number of women experiencing more than one or two deaths increases with an increase in the number of live-births.

It was further observed that comparatively much higher percentage of women experienced infant and child deaths when they had seven or more live-births as against those who had one to two or three to four live-births. A study conducted by the Population Research Centre, Kerala (1978) has also brought out the fact that for the larger family infant mortality is higher.

The relationship between number of live-births and number of deaths becomes still more clear when we look at the average number of live-births among those who had experienced child deaths and those who did not. Table 5.5 gives mean number of live-births for those who experienced child deaths and those who did not.

The table reveals significant variations in the mean number of live-births in the two categories for all the three villages as well as for each village separately. There was a difference of almost nearly two children between these two categories of women indicating thereby the close relationship between child mortality and fertility irrespective of the fact whether high mortality resulted in high fertility or vice-versa.

An inter-village comparison did not show any significant variation in the mean number of live-births among those respondents who experienced child deaths but among those who did not experience any child loss, mean number of live births was comparatively small in Ghudani Kalan.

The infant and child mortality has been found to vary with the order of birth which is a biological factor. Mortality rates are comparatively high at first birth order than at higher birth orders. Table 5.6 gives the incidence of child deaths by birth order.

The table shows that the highest percentage of deaths was observed for the first order birth while the lowest for the 5th

TABLE 5 5

Mean number of live-births among those who experienced child deaths and those who did not

Categories of respondents	Ugala		Ghagga		Ghudani Kalan		Total	
	Mean number of children born	Number & % of women	Mean number of children born	Number & % of women	Mean number of children born	Number & % of women	Mean number of children born	Number & % of women
Those who experienced child deaths	6.38	81 (35.06)	5.50	105 (40.86)	6.03	118 (40.55)	5.94	304 (39.02)
Those who did not	4.12	150 (64.94)	4.69	152 (59.14)	3.03	173 (59.45)	3.91	475 (60.98)
Total	4.91	231 (100.00)	5.02	257 (100.00)	4.26	291 (100.00)	4.70	779 (100.00)

TABLE 5.6

Infant and child deaths by birth/order

Order of birth	Ugala		Ghagga		Ghudani Kalan		Total	
	Number of deaths	% of deaths	Number of deaths	% of deaths	Number of deaths	% of deaths	Number of deaths	% of deaths
1	38	26.21	39	21.55	50	22.12	127	23.01
2	21	14.48	41	22.65	45	19.91	107	19.39
3	23	15.86	30	16.57	40	17.70	93	16.85
4	18	12.41	23	12.16	31	13.72	71	12.86
5	15	10.35	16	8.84	16	7.08	47	8.51
6 and more	30	20.69	33	18.23	44	19.47	107	19.39
Total	145	100.00	181	100.00	226	100.00	552	100.00

order births. The percentage of deaths again rises after 5th birth order for all the three villages as well as for each village. In fact, after the first birth order the child loss consistently declines till the 5th order where it is the lowest and then begin to rise. The Khanna Study (1971) which was carried out in 11 villages of Punjab, also brought out the fact that while the infant mortality rates were high for the first child, they were almost at the same level from the second to the sixth child and began to rise again only for the seventh and later children.

Further, when given the same order of birth, variations between the three villages were significant in the first two birth orders, whereas, in the remaining birth orders the differences were small. For example, the percentage of deaths for first birth order was comparatively higher for Ugala but for second birth order it was higher for Ghagga.

Maternal age is another important factor that influences infant deaths. On the basis of the data collected in the Greater Bombay fertility suruey, Ruzicka and Kanitkar (1972) found that the youngest and the oldest mothers experienced the highest infant and neo-natal mortality. Therefore, Table 5.7 gives the percentage of child deaths by mother's age.

In Table 5.7 the highest percentage of deaths were experienced by women in the age group of 15-20 years and then above 30 years of age in each village. The percentage of deaths was comparatively much less in the age group of 24-29 years. This trend, thus substantiates the findings of various other studies also.

It was further noted that given the same age groups, all the three villages exhibited more or less the same trend and inter-village variations were not significant.

In fact, maternal age and birth order are closely linked together as young mothers in earlier child births are mostly immature and inexperienced in this regard, and, therefore, experience greater child loss.

Child Mortality and Socio-Economic Factors

As already mentioned, the level of mortality is greatly influenced by various socio-economic factors. The low level of

TABLE 5.7

Infant and child death by mother's age

Age of the mother in years	Ugala		Ghagga		Ghudani Kalan		Total	
	Number of respondents	%	Number of respondents	%	Number of respondents	%	Number of respondents	%
15—20	40	27.58	45	24.86	65	28.76	150	27.18
21—23	31	21.38	30	16.57	36	15.93	97	17.57
24—26	17	11.72	25	13.81	34	15.04	76	13.77
27—29	15	10.35	25	13.81	30	13.27	70	12.68
30+	42	28.97	56	30.95	61	27.00	159	28.80
Total	145	100.00	181	100.00	226	100.00	552	100.00

literacy, low income, low level of medical facilities and unsatisfactory housing, are the major contributory factors to the high level of infant and child mortality. In the present analysis, the various socio-economic factors included are the educational level of the respondents and their husbands, their occupational and income status and their housing conditions.

Education is one of the most important factors in determining the level of infant and child mortality. The National sample survey in its 15th round came to the conclusion that mothers who had higher level of education had lower rate of infant mortality. It is because educated women have better knowledge of sanitation and medical facilities and are capable of better infant care. Table 5.8 gives the incidence of child deaths by the educational level of the respondents.

TABLE 5.8

Infant and child mortality by the educational level of the respondents

Educational level of the respondents	Ugala Number and % of children dead	Ghagga Number and % of children dead	Ghudani Kalan Number and % of children dead	Total Number and % of children dead
Illiterate	116 (12.61)	169 (14.23)	204 (20.08)	489 (15.65)
Literate	10 (14.08)	4 (12.50)	4 (13.33)	18 (13.53)
Primary	12 (13.95)	8 (17.78)	15 (12.30)	35 (13.83)
Middle	2 (8.33)	—	2 (8.00)	4 (7.27)
Secondary	5 (16.67)	—	1 (2.63)	6 (7.41)
Graduate	—	—	—	—
Total	145 (12.78)	181 (14.24)	226 (18.24)	552 (15.07)

Table 5.8 indicates the inverse relationship between the percentage of child deaths and the educational attainments of the respondents for all the three villages as well as for Ghudani Kalan but no such trend was observed in the other two villages. In Ugala the highest percentage of child deaths was observed among those who had been educated upto the secondary level. The incidence of child deaths was also higher for those respondents who were literate or had education upto primary level as compared to the illiterates. Similarly, in Ghagga child loss was higher for those who had education upto primary level and lower for the illiterates. Another notable fact was that in this village respondents beyond middle level of education did not experience any child loss. Thus, on the whole, except for Ghudani Kalan, there was no consistent trend in the relationship in the other two villages.

The variations in the percentage of children dead when given the same level of educational attainments of the respondents in these villages was also not very significant except for those who were illiterate or who had education upto the primary level.

Since the number of respondents who had been educated upto various levels, was quite small, therefore, much reliability cannot be placed on this variable. Hence, an attempt was made to relate the proportion of child loss to the educational attainments of the husbands. Also, because husbands educational level is an equally important factor determining child mortality. Table 5.9 gives the percentage of child deaths by the husbands educational attainments.

Table 5.9 reveals a more clear trend in the relationship between infant and child deaths and the educational level of the husbands. In all these villages, barring some inconsistencies in Ugala, the trend indicates inverse relationship between the two variables in Ghagga and Ghudani Kalan as well as in the total sample.

Another noticeable fact was that when given the same level of husband's educational attainments, the child loss was much higher in Ghudani Kalan as compared to the other two villages.

TABLE 5.9

Infant and child mortality by husband's educational level

Educational level of the husband	Ugala Number and % of children dead	Ghagga Number and % of children dead	Ghudani Kalan Number and % of children dead	Total Number and % of children dead
Illiterate	85 (14.78)	112 (15.41)	143 (21.22)	340 (17.20)
Literate	7 (8.64)	25 (13.59)	25 (20.49)	57 (14.73)
Primary	20 (13.42)	15 (12.30)	11 (16.42)	46 (13.61)
Middle	10 (7.19)	7 (9.59)	11 (13.25)	28 (9.49)
Secondary	10 (7.14)	10 (8.70)	16 (12.70)	36 (9.45)
Graduate	1 (7.69)	—	—	1 (2.04)
Not recorded	12 (31.58)	12 (18.75)	20 (14-93)	44 (18.64)
Total	145 (12.78)	181 (14.02)	226 (18.28)	552 (15.07)

The occupational status of the husband is another important factor that influences the level of child mortality. It is in fact the occupational status which determines the income, the residence and all other things in the life of a person. Table 5.10 gives the occupational status of the husbands and the percentage of child deaths.

As already stated, the occupations followed by the husbands were ranked into seven prestige categories and in the given table the various occupations are represented by these categories. The table clearly reveals the inverse relationship between the percentage of child loss and the occupational status of the husband. The percentage of child deaths was the lowest where husbands were engaged in high prestige occupational category

TABLE 5.10

Infant and child mortality by the occupational status of the husband

Occupational prestige categories	Ugala Number and % of children dead	Ghagga Number and % of children dead	Ghudani Kalan Number and % of children dead	Total Number and % of children dead
I—III	8 (7.34)	5 (6.25)	3 (13.64)	16 (7.58)
IV—V	54 (10.31)	73 (12.03)	104 (17.33)	231 (13.34)
VI—VII	62 (14.59)	77 (15.62)	71 (22.76)	210 (17.07)
No occupation	21 (21.27)	26 (23.42)	48 (15.90)	95 (19.39)
Total	145 (12.78)	181 (14.02)	226 (18.28)	552 (15.07)

one to three and the percentage of child deaths increases consistently with the lowering of occupational prestige and was much higher where the husbands were employed in low prestige occupations, or, where they had no occupation in the total sample as well as in each village.

An inter-village comparison reveals that when given the same occupational status of the husband, the percentage of child loss was much higher, except in the last category where the husband's occupation was not specified, in Ghudani Kalan in each occupational category as compared to other two villages.

The income of the family also plays a significant role in determining the level of infant and child mortality. It is because, apart from other advantages, people belonging to higher socio-economic strata have an easy access to the available medical facilities. However, the data in the present study did not show any correlation between the income level and extent of infant and child deaths in the total sample of the population as well as in each village individually, therefore, it was not presented in the form of a table.

Another indicator of the economic status is the level of living of the family. No doubt, income of the family is reflected in the level of living but, since it is a combination of various factors, it might give us a better idea of the relationship between the economic status and the extent of child mortality. Table 5.11 gives the distribution of respondents by their level of living and the percentages of child loss.

Table 5.11 indicates inverse correlation between these two variables. On the whole, the proportion of child loss was much higher where the standard of living was low, and lower, where it was high in all the three villages as well as in each village separately and the difference in mortality between the high and the low level of living was quite significant.

An inter-village comparison reveals that at each level of living the proportion of child loss was comparatively much higher in Ghudani Kalan and lower in Ugala.

Lastly, housing conditions are another important factor related to the incidence of child mortality. Chandrasekhar (1972) also pointed out that the infant mortality rate ranged in

TABLE 5.11

Infant and child mortality by level of living

Level of living	Ugala Number and % of children dead	Ghagga Number and % of children dead	Ghudani Kalan Number and % of children dead	Total Number and % of children dead
Low	78 (16.49)	64 (17.63)	57 (23.65)	199 (18.48)
Medium	27 (11.12)	69 (13.91)	50 (19.60)	146 (14.70)
High	40 (9.52)	48 (8.82)	119 (16.08)	207 (13.00)
Total	145 (12.78)	181 (14.02)	226 (18.28)	552 (15.07)

magnitude according to the area of residence, type of dwelling and the availability of basic amenities, etc. In Mysore Study (1961), it was revealed that the infant mortality rate for families in poor-class housing was 100 per 1,000 live-births but the rate dropped to 58 when the housing conditions improved.

Table 5.12 gives the percentage of infant deaths by the type of dwelling.

TABLE 5.12

Infant and child mortality by the type of dwelling

Type of residence	Ugala Number and % of children dead	Ghagga Number and % of children dead	Ghudani Kalan Number and % of children dead	Total Number and % of children dead
Permanent	56 (11.31)	100 (12.03)	103 (15.97)	259 (13.14)
Semi-permanent	31 (11.31)	24 (15.69)	88 (19.69)	143 (16.36)
Temporary	58 (15.85)	57 (18.57)	35 (24.31)	150 (18.36)
Total	145 (12.78)	181 (14.02)	226 (18.28)	551 (15.07)

The data in the above table clearly reveals an inverse relationship between the type of dwelling and the percentage of child deaths in the total sample as well as for each village, and the difference between the houses made of bricks and concrete and the mud huts was quite significant for all the villages. This may be because along with the type of house go other factors like drainage, garbage disposal, running water and general sanitation.

Another important point to be taken note of in the above table is that when given the same type of dwelling unit in each village, the percentage of infant and child deaths was considerably higher in Ghudani Kalan as compared to other two villages.

Thus an overview of the above analysis indicates that, on the whole, the general level of mortality was comparatively higher for Ghudani Kalan and lower for Ugala while Ghagga stands at the middle level. Similar trend was observed in infant and child deaths. The data also reveal age and sex differentials in infant and child mortality as more deaths occurred in the neo-natal period as compared to post neo-natal and the age group of 1-4 years. It was further noted that more male deaths occurred in the neo-natal stage whereas, more female deaths occurred in the post neo-natal and later period.

Infant and child mortality was closely related to fertility as the child loss was much higher for those respondents who had higher number of live-births. The percentage of child deaths was also related to the birth order and age of the mother. Lastly, the incidence of infant and child mortality was found to vary with the socio-economic level of the respondents.

REFERENCES

1. Chandrasekhar, S. 1972: Infant Mortality, Population Growth and Family Planning in India, London, George Allen and Unwin.
2. Devi, V.S. 1978: 'Effective Perception of Infant Mortality on Actual Family Size', The Journal of Family Welfare, Vol. xxiv, 4, June.
3. Jain, S.P. 1979: Levels and Differentials of Infant and Child Mortality; Determinants and Demographic Impact, in Srinivasan, K. et al. (eds.) Demographic and Socio-Economic Aspects of the Child in India, Bombay, Himalaya Publication House.
4. Ruzicka, L.T. and Kanitkar, Tara 1972: Infant Mortality in an Urban Setting. The case of Greater Bombay, in Vaidyanathan (ed.) Studies on Mortality in India, Gandhigram, Institute of Rural Health and Family Planning.
5. Singha, P. 1975: Infant Mortality and the Level of Fertility in India: A Review, Demography India, Vol. 4.2.
6. United Nations 1973: The Determinants and Consequences of Population Trends, New Summary of Findings on Interaction of Demographic, Economic and Social Factors, Population Studies, No. 50, New York.
7. United Nations 1961: The Mysore Population Study, New York.
8. Wyon John W. and Gordon John E. 1971: The Khanna Study: Population Problems in the Rural Panjab, Cambridge, Harvard University Press.

6

SOCIO-ECONOMIC STATUS AND FERTILITY

In the present chapter an attempt has been made to examine some of the socio-economic determinants of fertility behaviour in these rural areas. The variables included under socio-economic determinants are education, occupation, income, level of living, land ownership and housing conditions. It has been universally accepted that socio-economic changes depress higher fertility among different groups, therefore, the present chapter is an attempt to see whether there are any differentials in fertility due to different socio-economic conditions in these rural areas.

Education and Fertility

Education, especially, of women is considered to be one of the most important variables affecting fertility behaviour. Berelson, while reviewing the recent literature observes, "educational attainment has a consistently inverse relationship to fertility in almost all instances, and is moreover, one of the strongest relationship between a stratification variable and fertility. Similarly, the inverse relationship of education to completed family size is one of the most clear cut correlations found in the literature (Berelson 1976)."

Education affects fertility through the value of two intermediate variables, viz., age at marriage, as well as, in the attitude regarding the number of children desired and in the acceptance of family planning methods. Generally higher educated people have higher age at marriage and higher age at marriage reduces the total reproductive span. Moreover, higher education, especially, among women is one of the indicators of modernization and higher social status which, in turn, influences the

TABLE 6.1

Educational level of the respondents Vs. mean number of live-births

Educational level	Ugala		Ghagga		Ghudani Kalan		Total	
	Mean no. of live-births	No. & % of respondents	Mean no. of live-births	No. & % of respondents	Mean no. of live-births	No. & % of respondents	Mean no. of live-births	No. & % of respondents
Illiterate	5.20	177 (75.32)	4.96	240 (87.91)	4.65	219 (73.00)	4.94	636 (78.71)
Literate	4.43	16 (6.81)	4.00	8 (2.93)	3.33	9 (3.00)	4.15	33 (4.08)
Primary	4.10	21 (8.93)	4.00	12 (4.40)	3.29	37 (12.33)	3.65	70 (8.66)
Middle	3.29	7 (2.98)	1.50	4 (1.47)	2.27	11 (3.67)	2.45	22 (2.72)
Secondary	2.50	12 (5.11)	1.75	7 (2.56)	1.90	20 (6.67)	2.06	39 (4.84)
Graduate and above	2.00	2 (0.84)	1.25	2 (0.73)	1.25	4 (1.33)	1.44	8 (0.99)
Total	4.02	235 (100.00)	4.73	273 (100.00)	4.12	300 (100.00)	4.53	808 (100.00)

behaviour and attitude towards family size. According to most of the studies conducted in India, the effect of education in depressing fertility is visible only at higher educational level that is beyond the school level. In Bangalore city (UN 1961), women aged 45 years and over who had ever been married and had a high school or University education had 4.0 children compared to 5.4 for those with less education.

The table clearly indicates negative association between the level of educational achievements and the mean number of live-births for all the villages as well as for each village separately. Although the number of women at each educational level was rather small, still given the number of women at each educational level, mean number of live-births was the highest among illiterate respondents and the number shows consistent decline with an increase in the educational level of the respondents for all the villages as well as for each village separately. The data also indicate that when given the same level of educational attainments in each village, mean number of live-births was comparatively higher for Ugala and little lower for Ghudani Kalan.

As pointed out earlier in the chapter on age at marriage, higher educated respondents had much higher age at marriage and there was a difference of 5.7 years in the mean age at marriage between higher educated and less educated or uneducated respondents, therefore, this variation in the mean number of live-birth between educated and uneducated respondents could partly be due to their higher age at marriage. This fact was further substantiated when it was controlled for marriage duration.

Since the proportion of women who were literate or who were educated upto various levels was not very large, it did not appear to be very safe to depend upon this variable entirely as a determinant of fertility behaviour. Therefore, an attempt had been made to relate number of live-births to the educational level of the husbands. Moreover, husband's educational level has also been found to have depressing effect on fertility. Khanna Study (1971) had revealed dissimilarity in birth rate by the education of husband. Driver (1970) also found such differentials in the study of Differential Fertility in Central India.

Table 6.2 gives the distribution of respondents by their husband's educational level and their mean number of live-births.

In Table 6.2, barring the category of literate husbands in Ghagga and Ghudani Kalan, there was a clear inverse relationship between husband's educational attainments and mean number of live-births in each village as well as for all of them. Only in these two villages mean number of live-births were more for literate husbands as against illiterate ones. But mere literacy does not mean enough education, therefore, no generalization could be made for that. On the whole, from primary level onwards, mean number of live-births consistently declined with an increase in the educational level of husbands.

However, an inter-village comparison reveals that when given the same level of educational attainments, with few exceptions, mean number of live-births was higher for Ugala and lower for Ghudani Kalan and Ghagga falls at the middle level.

Female Labour Force Participation and Fertility

Female participation in the labour force has often been considered as one of the important variables that influences fertility. The Mysore Population Study (UN 1961) suggests a definite association between fertility and the working status of women. The average number of live-births is systematically lower in all age groups and strata for gainfully occupied women. It has been further found that women's employment where unaccompanied by education, does not have a significant effect in lowering fertility. Thus, it is not female labour force participation per se that determines fertility behaviour but the effect of work on fertility depends on the work setting, type of work, role incompatibility and educational level of the women.

Unfortunately, the number of gainfully employed women was very small in the present study and this trend appeared to be in accordance with the trends in the states of Punjab and

TABLE 6.2

Educational level of husbands Vs. mean number of live-births

Educational level of the husband	Ugala		Ghagga		Ghudani Kalan		Total	
	Mean no. of live-births	No. & % of husband	Mean no. of live-births	No. & % of husband	Mean no. of live-births	No. & % of husband	Mean no. of live-births	No. & % of husband
Illiterate	5.46	105 (44.68)	5.13	142 (52.01)	4.85	139 (46.33)	5.12	386 (47.77)
Literate	5.40	15 (6.38)	6.34	29 (10.62)	5.08	24 (8.00)	5.69	68 (8.42)
Primary	5.31	28 (11.92)	4.06	30 (10.90)	3.35	20 (6.67)	4.33	78 (9.65)
Middle	4.18	33 (14.04)	3.43	21 (7.69)	3.32	25 (8.33)	3.71	79 (9.78)
Secondary	3.23	44 (18.72)	3.19	36 (13.19)	2.74	46 (15.34)	3.04	126 (15.59)
Graduate & above	2.60	5 (2.13)	1.50	4 (1.47)	2.50	12 (4.00)	2.33	21 (2.60)
Not recorded	7.40	5 (2.13)	5.81	11 (4.03)	4.00	34 (11.33)	4.74	50 (6.13)
Total	4.82	235 (100.00)	4.73	273 (100.00)	4.12	300 (100.00)	4.53	808 (100.00)

Haryana where the percentage of women workers was the lowest in the country. However, among those who were working, majority of them were engaged in agriculture and it has often been found that the fertility of women engaged in agricultural activities or in household industry, is higher, or, does not differ significantly from that of non-working women. Tables 6.3 and 6.4 gives the mean number of live-births among working and non-working and by their employment status.

Table 6.3 shows that for all the three villages as well as for Ghudani Kalan, working women had lesser number of live-births as compared with non-working ones, although the difference was very small for all of them but significant for Ghudani Kalan. In the other two villages, non-working women had lesser number of live-births than the working women. A comparative picture of the villages reveals that among non-working women, there were not well marked differences in the mean number of live-births among these villages, but these differences were very significant among working women and fertility level among working women was the highest in Ugala and the lowest in Ghudani Kalan.

These differences indicate differences in the work force status of the women in these villages. In Ugala, where mean number of live-births was the highest, out of four employed women, three were working as agricultural labourers. Similarly, in Ghagga, of the 31 working women, 25 were working as agricultural labourers, whereas, in Ghudani Kalan, none of them was working as an agricultural labourer and majority of them were school teachers or doctors, and therefore, they had lesser number of live-births. Thus, on the whole, the differentials by the status of employment were clear for each villages.

The occupational status of the husband is another important variable that influences fertility. The Mysore Study (UN 1961) has revealed that in rural areas, completed family size of agricultural labour was 4.0, of small cultivators 4.7 and of large cultivators 5.1, but Driver's Study (1960) has not revealed any significant variations. Table 6.5 gives the distribution of respondents by their husband's occupational status and mean number of live-births born to them.

TABLE 6.3

Mean number of live-births among working and non-working women

Categories of women	Ugala		Ghagga		Ghudani Kalan		Total	
	Mean no. of live-births	No. & % of respon-dents	Mean no. of live-births	No. & % of respon-dents	Mean no. of live-births	No. & % of respon-dents	Mean no. of live-births	No. & % of respon-dents
Non-working women	4.80	231 (98.30)	4.69	242 (88.64)	4.20	291 (97.00)	4.54	764 (94.55)
Working women	6.25	4 (1.70)	4.90	31 (11.36)	2.00	9 (3.00)	4.43	44 (5.45)
Total	4.82	235 (100.00)	4.73	273 (100.00)	4.12	300 (100.00)	4.53	808 (100.00)

TABLE 6.4

Occupational status of the respondents Vs. mean number of live-births

Occupational status of women	Ugala		Ghagga		Ghudani Kalan		Total	
	Mean no. of live-births	No. & % of respondents	Mean no. of live-births	No. & % of respondents	Mean no. of live-births	No. & % of respondents	Mean no. of live-births	No. & % of respondents
Doctor	—	—	—	—	1.00	1 (11.11)	1.00	1 (2.27)
Teacher	5.00	1 (25.00)	3.00	2 (6.45)	1.40	5 (55.56)	2.25	8 (18.18)
Small shopkeeper	—	—	4.00	1 (3.23)	—	—	2.25	1 (2.27)
Small farmer	—	—	6.50	2 (6.45)	—	—	6.50	2 (4.55)
Skilled labourer	—	—	1.00	1 (3.23)	1.00	1 (11.11)	1.00	2 (4.55)
Domestic/Shop servant	—	—	—	—	3.00	1 (11.11)	3.00	1 (2.27)
Sweeper	—	—	—	—	7.00	1 (11.11)	7.00	1 (2.27)
Agricultural labourer	6.66	3 (75.00)	5.12	25 (80.64)	—	—	5.29	28 (63.64)
Total	6.25	4 (100.00)	4.90	31 (100.00)	2.11	9 (100.00)	4.45	44 (100.00)

TABLE 6.5

Occupational status of the husband Vs. mean number of live-births

Occupational prestige categories	Ugala		Ghagga		Ghudani Kalan		Total	
	Mean no. of live-births	No. & % of respondents	Mean no. of live-births	No. & % of respondents	Mean no. of live-births	No. & % of respondents	Mean no. of live-births	No. & % of respondents
(1)	(2)	(3)	(4)	(5)	(6)	(7)	(8)	(9)
1	—	—	—	—	3.00	1 (0.33)	3.00	1 (0.12)
2	4.20	15 (6.38)	4.62	8 (2.93)	2.00	1 (0.33)	4.25	24 (2.97)
3	5.75	8 (3.40)	5.37	8 (2.93)	4.25	4 (1.34)	5.30	20 (2.48)
4	4.77	50 (21.28)	4.35	42 (15.38)	3.26	45 (15.00)	4.15	137 (16.96)
5	4.15	69 (29.36)	4.37	97 (35.53)	3.72	122 (40.67)	4.04	288 (35.63)

(*Contd.*)

TABLE 6.5—*Contd.*

(1)	(2)	(3)	(4)	(5)	(6)	(7)	(8)	(9)
6	3.99	17 (7.24)	4.30	23 (8.43)	4.18	21 (7.00)	4.17	61 (7.55)
7	5.45	65 (27.66)	5.12	77 (28.21)	5.23	43 (14.33)	5.26	185 (22.90)
No occupation	7.00	11 (4.68)	6.22	18 (6.59)	5.62	29 (9.67)	6.07	58 (7.18)
Information not available	—	—	—	—	4.15	34 (11.33)	4.15	34 (4.21)
Total	4.82	235 (100.)	4.73	273 (100.)	4.12	300 (100.)	4.53	808 (100.)

As mentioned earlier, all the occupations followed by the husbands had been ranked into seven prestige categories and category one represents occupations of the highest prestige, whereas category seven represents occupations of the lowest prestige.

Table 6.5 indicates that although these respondents whose husbands were engaged in the lowest category of occupation had the highest number of live-births but reverse was not true for respondents whose husbands were engaged in higher prestige occupations. Further, in between the various occupational categories also, there was no consistency, showing, thereby, lack of any association between these two variables.

An inter-village comparison also did not show any uniform pattern when given the same category of occupations.

Economic Status and Fertility

It is generally held that there is an inverse correlation between economic prosperity and fertility. Demographers explain this by saying that at the higher levels of prosperity one gets greater amenities of life and greater pleasure seeking opportunities, and, as a consequence, the importance of sex declines. But wealth according to Thompson (1965), in itself, has little influence on birth rates. It is the whole set of social circumstances, living conditions and social relationships that accompany wealth that is really the important factor. In the present study, under economic status has been included the actual family income, level of living as well as land ownership. Table 6.6 gives the distribution of respondents by their family income and mean number of live-births.

The data in Table 6 6 do not reveal any clear association between income and the mean number of live-births in the total as well as in each village as there was no consistency between various income categories, although, in some cases mean number of live-births was higher for lower income groups and vice-versa. It also revealed no uniform trend in the mean number of live-births when given the same income level of the respondents between the villages. Jain (1975) while discussing the relationship between fertility and income rightly concludes that fertility differentials due to income are small especially in

TABLE 6.6

Family income Vs. mean number of live-births

Income categories	Ugala		Ghagga		Ghudani Kalan		Total	
	Mean no. of live-births	No. & % of respondents	Mean no. of live-births	No. & % of respondents	Mean no. of live-births	No. & % of respondents	Mean no. of live-births	No. & % of respondents
Less than Rs. 2,500	4.47	44 (18.72)	4.51	58 (21.25)	5.00	39 (13.00)	4.63	141 (17.45)
Rs. 2,500 to 5,000	5.15	46 (19.57)	4.60	52 (19.05)	4.15	77 (25.67)	4.55	175 (21.66)
Rs. 5,000 to 7,500	5.28	39 (16.60)	5.50	60 (21.98)	4.04	59 (19.67)	4.90	158 (19.56)
Rs. 7,500 to 10,000	4.58	17 (7.23)	4.10	32 (11.72)	4.20	25 (8.33)	4.24	74 (9.16)
Rs. 10,000 to 15,000	4.37	32 (13.62)	4.60	41 (15.02)	4.26	38 (12.67)	4.42	111 (13.74)

Rs. 15,000 to 20,000	4.67	28 (11.92)	4.15	13 (4.76)	4.33	9 (3.00)	4.47	50 (6.19)
Rs. 20,000 to 30,000	4.91	22 (9.36)	3.75	8 (2.93)	3.42	19 (6.33)	4.14	49 (6.06)
Rs. 30,000 and above	4.50	6 (2.55)	4.50	4 (1.46)	3.24	33 (11.00)	3.53	43 (5.32)
Not known	5.00	1 (0.43)	6.60	5 (1.83)	0.1	1 (0.33)	5.44	7 (0.86)
Total	4.82	235 (100.00)	4.73	273 (100.00)	4.12	300 (100.00)	4.53	800 (100.00)

rural areas and, possibly, it may be because cultural practices, which are the real determinants of fertility behaviour in India, do not differ with change in income.

Level of living is another economic variable that can help us in understanding better the relationship between fertility and economic status as it includes various other economic aspects besides income. As mentioned in the methodology, level of living index was prepared and divided into three categories, namely, high, medium and low, to gauge the economic status of the respondents. Therefore, Table 6.7 gives the mean number of live-births by the level of living index.

In Table 6.7 there is negative association between the level of living and mean number of live-births in the total as well as in Ugala and Ghudani Kalan, whereas, in Ghagga the highest number of live-births were observed at the middle level of living and statistically this relationship was significant at 0.05 level.

An inter-village comparison indicates that at the two extreme ends, i.e., at the higher and also at the lower level of living, mean number of live-births were found to be higher for Ugala and lower for Ghudani Kalan and Ghagga remained at the middle level. But at the middle level of living mean number of live-births was higher for Ghagga followed by Ugala and Ghudani Kalan. Thus level of living appears to be a better index of economic status as compared with simple income categories.

Since it is a study of rural areas and the major source of income in rural areas being landed property or the wages earned from agriculture labour, it would be interesting to see if there are any variations in fertility among those who owned land and those who did not. and among those who owned land by the size of the land holdings. Driver (1963) in his study found 4.5 as average number of children for the landless as against 4.6 for the agriculturists. And among land owners, the average family size was slightly higher for large land owners as compared to small land owners. Mamdani (1972) while commenting on Khanna Study stated that in villages every one seemed to want more rather than less children. The landed wanted more children, especially sons, to work the family farm more efficiently and dispense with expensive hired labour and if possible, increase

TABLE 6.7

Respondent's level of living Vs. mean number of live-births

Level of living	Ugala		Ghagga		Ghudani Kalan		Total	
	Mean number of live-births	Number & % of respondents	Mean number of live-births	Number & % of respondents	Mean number of live-births	Number & % of respondents	Mean number of live-births	Number & % of respondents
Low level	5.25	90 (38.30)	4.46	81 (29.67)	4.82	50 (16.67)	4.86	221 (27.35)
Medium level	4.64	52 (22.13)	5.38	92 (33.70)	4.06	63 (21.00)	4.79	207 (25.62)
High level	4.50	93 (39.57)	4.32	100 (36.63)	3.96	187 (62.33)	4.18	380 (47.03)
Total	4.82	235 (100.)	4.73	273 (100.)	4.12	300 (100.)	4.53	808 (100.)

the holding in course of time. The landless wanted several children because the only hope of raising themselves from their wretched condition lay in having many sons who would find employment and add to family earnings and, possibly, enable the family to acquire a patch of land. The artisans, too, had the same logic. Srinivas (1977) considers Mamdani's views as very optimistic and thinks that the number of children that a family can usefully absorb on the farm depends on the size of its land holdings. He further says that the question of partitioning the land among the sons may also be taken into account. The more the sons, the smaller the share of each in the family holdings.

In the present sample little more than fifty per cent of the respondents owned land although this proportion was different for each village. However, the percentage of land owners was the highest (61.00) in Ghudani Kalan followed by Ugala (58.00) and the lowest (40.00) in Ghagga. Table 6.8 gives mean number of live-births by those who owned land and those who did not.

Table 6.8 reveals that in every village land owners had lesser number of live-births as against those who did not own any land and the difference between the two categories of respondents was comparatively larger in Ghudani Kalan and smaller in Ugala. Another important point noted was that among respondents of both the categories, whether land owners, or those who did not own any land, number of live-births was comparatively higher in Ugala and lower in Ghudani Kalan and Ghagga falls at the middle level. Moreover, the difference in the number of live-births was more significant between Ugala and Ghudani Kalan rather than Ugala and Ghagga or Ghagga and Ghudani Kalan.

Thus, on the whole, the land owners had lesser number of live-births than the non-land owners and this may be because farming has become more mechanised now and the farmers use many labour saving devices, therefore, they do not depend much on family labour. This relationship perhaps would become more clear if we examine the size of the land holdings. It is also because the number of children that a family can usefully absorb on its farm depends on the size of its land holdings.

In the present study large majority of the farmers had small land holdings, i.e., between 1 to 10 acres and nearly one fifth of

TABLE 6.8

Mean number of live-births among those who owned land and those who did not

	Ugala		Ghagga		Ghudani Kalan		Total	
	Mean number of live-births	Number & % of respondents	Mean number of live-births	Number & % of respondents	Mean number of live-births	Number & % of respondents	Mean number of live-births	Number & % of respondents
Those who did not own any land	4.89	99 (42.13)	4.78	165 (60.44)	4.45	118 (39.33)	4.71	382 (47.28)
Land owners	4.80	135 (57.87)	4.64	108 (39.56)	3.92	182 (60.67)	4.38	426 (52.72)
Total	4.82	235 (100.)	4.73	273 (100.)	4.12	300 (100.)	4.53	808 (100.)

them had 10-15 acres of land and the number of farmers owning more than 15 acres of land was quite small. Table 6.9 gives the distribution of respondents by the size of their land holdings and mean number of live-births.

Table 6.9 shows that mean number of live-births varies inversly with the size of the land holdings for the total sample as also for Ghudani Kalan. However, in Ugala and Ghagga medium farmers had higher number of live-births as compared to the other two categories. But, comparatively, farmers with big land holdings had lesser number of live-births.

Besides, more number of live-births, it is also important to look at the number of living sons which farmers of each category have as a large land holding can be reduced to several marginal land holdings in due course of time. But here, too, the trend was not clear in all the villages, although farmers with medium land holdings had comparatively higher number of living sons.

Thus, on the whole farmers with big land holdings had not only smaller number of live-births but also smaller number of living sons and division of land holdings might have been a consideration for them in having smaller family size but number of such cases is rather too small to justify making any such generalization. Wyon and Gordon (1971) also indicated that this could be a cause for serious concern but they also did not elaborate it.

Housing Conditions and Fertility

Type of dwelling is also an index of the economic status of the family. The economic conditions, in fact, determine the type of residence as well as the locality where one resides. In India there are not many studies relating family size with the type of dwelling. Mysore Study (UN 1961) revealed a positive association between upper status and completed family size as it was noted that hut dwellers had an average of 4.4 children, dwellers in mud houses with thatched roof had 4.5 children and dwellers in still better houses had 5.0 children. In the present study the dwelling unit was divided into three categories, namely, permanent, semi-permanent and temporary representing

TABLE 6.9

Size of land holding Vs. mean number of live-births

Size of land holdings	Ugala		Ghagga		Ghudani Kalan		Total	
	Mean number of live-births	Number & % of respondents	Mean number of live-births	Number & % of respondents	Mean number of live-births	Number & % of respondents	Mean number of live-births	Number & % of respondents
1—10 acres	4.72	84 (61.76)	4.68	76 (70.37)	4.23	130 (71.43)	4.49	290 (68.08)
10—25 acres	4.94	39 (28.68)	4.92	26 (24.07)	3.17	42 (23.08)	4.24	107 (25.12)
25 acres and more	4.89	13 (9.56)	2.87	6 (5.56)	3.10	10 (5.49)	3.86	29 (6.80)
Total	4.80	136 (100.)	4.64	108 (100.)	3.82	182 (100.)	4.38	426 (100.)

Pakka, Semi-Pakka and Kacha houses. Table 6.10 gives the mean number of live-births by the type of dwelling.

The table shows close association between the type of dwelling and the mean number of live-births in the total as well as in Ghudani Kalan but there was no clear trend for the other two villages. For instance, in Ugala, although number of live-births was the lowest among respondents living the Pakka houses made of bricks and concrete but the highest number of live-births was not found among hut dwellers. Similar discrepancies were found in Ghagga. An inter-village comparison also did not reveal any uniform trend as among respondents who lived in Pakka houses, number of live-births was the highest in Ghagga, but among those who lived in semi-pakka houses, number of live-births was the highest in Ugala whereas among hut dwellers it was the highest in Ghudani Kalan.

Thus, on the whole the influence of dwelling place was there for the total sample but not for any village taken by itself.

Migration and Fertility

It is often stated that the behaviour of a migrant may be different from that of the resident population. But the specific nature of the association between migration and fertility depends upon the characteristics of the migrants, place of migration as well as their perception of costs and returns on the part of the parents.

A review of the studies (Bhatia and Sabagh 1980) which have looked into the relationship between migration and fertility reveals inconsistency in the findings of the various studies. While some studies have reported higher fertility for the migrants than the non-migrants at the place of destination, others have shown the fertility of natives to exceed that of the migrants.

Many studies have shown that since migrants to urban areas are more active, dynamic, educated and belong to higher socio-economic level, they probably exhibit lower fertility (Bulsara, Jal F. 1964). But for urban to rural migration Bhatia and Sabagh (1980) assumed that they not only manifest lower fertility themselves, but may also diffuse the lower fertility norms in the rural areas, and thus prove as catalysts of change. Most of these studies have taken into account either rural to urban migration

TABLE 6.10

Type of house Vs. mean number of live-births

Dwelling type	Ugala		Ghagga		Ghudani Kalan		Total	
	Mean number of live-births	Number & % of respondents	Mean number of live-births	Number & % of respondents	Mean number of live-births	Number & % of respondents	Mean number of live-births	Number & % of respondents
Pakka	4.34	114 (48.51)	4.78	174 (63.74)	3.98	162 (54.00)	4.38	450 (55.69)
Semi-Pakka	5.96	46 (19.57)	4.50	34 (12.45)	3.99	112 (37.33)	4.55	192 (23.76)
Kacha	4.88	75 (31.92)	4.72	65 (23.81)	5.54	26 (8.67)	4.92	166 (20.55)
Total	4.82	235 (100.)	4.73	273 (100.)	4.12	300 (100.)	4.53	808 (100.)

or urban to rural migration. But there are hardly any studies to indicate the impact of rural to rural migration on fertility. In the present study 24 per cent of the sample population consisted of migrants and majority of them had migrated from one rural area to another. Therefore, it would be interesting to see whether there are any fertility variations among migrants and non-migrants. Table 6.11 gives the mean number of live-births among migrants and non-migrants.

The table indicates that the total migrants had higher number of live-births as compared with non-migrants. However, there was no uniform trend visible for all the three villages. In Ugala and Ghudani Kalan migrants had higher number of live-births, whereas, in Ghagga, non-migrants had higher number of live-births. Thus, on the whole, these differences were there but they were not well marked, therefore, no generalization could be made on the basis of these small differences.

But when these data were analysed with duration of marriage, it was found that in all the three villages, with longer duration of marriage, i.e., 20 years and more, non-migrants had higher number of live-births than the migrants.

Thus, on the whole, there were no significant differences in the mean number of live-births between migrants and non-migrants and this may be because these were mostly rural to rural migrants not exposed to urban culture and values.

It may thus be summarised from the above that educational level of the respondents as well as that of their husbands was negatively correlated with live-births. Working women, on the whole, had lesser number of live-births than the non-working but among working women, those engaged in agricultural labour and other lower jobs had much higher number of live-births as against those who were educated and working as teachers or in other higher occupations. Husband's occupational status did not show any correlation with number of live-births.

Family income did not reveal any correlation with the number of live-births. But number of live-births varied among land owners and those who did not own any land and also with the size of land holdings. Further, no significant variations were observed in the mean number of live-births among migrants and non-migrants. Besides these intra-village variations, inter-village

TABLE 6.11

Mean number of live-births among migrants and non-migrants

Migrants/ Natives	Ugala		Ghagga		Ghudani Kalan		Total	
	Mean number of live-births	Number & % of respondents	Mean number of live-births	Number & % of respondents	Mean number of live-births	Number & % of respondents	Mean number of live-births	Number & % of respondents
Migrants	4.81	85 (36.17)	4.69	97 (35.53)	4.20	10 (3.33)	4.73	192 (23.76)
Non-migrants	4.83	150 (63.83)	4.72	173 (63.37)	4.12	290 (96.97)	4.46	613 (75.87)
Not recorded	—	—	6.33	3 (1.10)	—	—	6.33	3 (0.37)
Total	4.82	235 (100.)	4.73	273 (100.)	4.12	300 (100.)	4.53	808 (100.)

variations in the mean number of live-births when given the same socio-economic status, were also substantial.

REFERENCES

1. Berelson, Bernard 1976: "Social Science Research on Population", Population and Development Review, Vol. 2, 2, June.
2. Bulsara, Jal F. 1964: Problems of Rapid Urbanization in India, Bombay, Popular Prakashan.
3. Driver, E.D. 1960: "Fertility Differentials among Economic Strata in Central India, Eugenic Quarterly, Vol. 7, 2, June.
4. Hussain, I.Z. 1970: "Educational Status and Differential Fertility in India", Social Biology, Vol. 17.
5. Jain, S.P. 1975: "Demography—A Status Study on Population Research in India, Vol. II, New Delhi, Tata McGraw-Hill Publishers.
6. Jagdish, C. Bhatia and Georges Sabagh 1980: "Migration and Fertility in India", Demography India, Vol. 9, Nos. 1&2, Jan-Dec.
7. Mamdani, M. 1972: The Myth of Population Control, London, Monthly Review Press.
8. Nayar, P.K.B. 1974: "The Influence of Education on Fertility", the Journal of Family Welfare, Vol. XX, 3.
9. Singh K.P. 1979: Status of Women and Population Growth in India, New Delhi, Munshiram Manoharlal Publishers.
10. Srinivas, M.M. and Ramaswamy, E.A. 1977: Culture and Human Fertility in India, New Delhi, Oxford University.
11. Thompson, W.S. and Lewis, D.T. 1965: Population Problems, New York, McGraw-Hill Publishers.
12. United Nations 1961: Mysore Population Study, New York.

7

CASTE STATUS AND FERTILITY

HINDU society has been historically stratified on the basis of caste and an individual's behaviour is, to a great extent, determined by the caste norms. Today, in spite of the recent changes and the process of modernization since India's Independence, caste structure has not undergone much change and the general behaviour of people belonging to different caste groups has not altered significantly.

Caste as a stratification variable has not been studied systematically in studies of differential fertility. The Indian census ceased to collect data by caste since 1941, therefore, we cannot get all-India picture. In the absence of any national level study, it is difficult to understand the mechanisms through which caste affects the growth of population. However, a few local surveys which incorporated this variable to explain the differences in fertility are the Mysore Population Study (UN 1961), Driver (1963), Rele (1963), Saxena (1965) and Khanna Study (1971).

While studying the effects of caste on fertility in the rural areas, Rele (1963) reported that upper class Hindus, consisting of Brahmins and Kshatriya castes, had the lowest fertility and the Muslims had the highest. Sexena (1965) also observed that upper caste women had fewer than 7.3 live-births by the end of their reproductive period as compared with 8 and 8.3 living births for intermediate and low castes, respectively. Analysing the data from Khanna Study (1971), Gordon et al. noticed that mean number of live-births among Jat women of age 45 years and over was 6.78, whereas, it was 8.3 for chamar women. On the whole, the general conclusion is that lower castes like sweepers and chamars have higher fertility than the other castes.

Most of these studies do not indicate whether it is caste per se which contributes to these differentials in ferility, or, there are other intervening variables which also contribute to these variations. In most of these studies only one variable has been taken into consideration for explaining the differentials in fertility. But the relationship between caste and fertility appears to be a complex one as many variables of socio-economic and demographic background also influence fertility and the influence of caste may operate indirectly through these variables rather than directly.

Therefore, the present chapter is an attempt to examine the relationship between the caste status of the respondents and their fertility behaviour. It is assumed that caste status per se is not an important factor that influences fertility. Caste influences fertility through the intervening variables of age at marriage, the extent of child loss and the socio-economic conditions of the respondents.

For examining this relationship, all the castes in the three villages were divided into four categories. Category one includes all the higher castes such as Brahmins, Kshatriyas and Rajputs. Castes at the middle level like Vaishyas, Aroras and Sainis etc., were included in category two. In the third category have been included all those castes which have been declared as backward by the Government of Punjab and Haryana and category four consisted of scheduled castes. Table 7.1 gives the distribution of respondents according to their caste status and mean number of live-births.

In Table 7.1 there is clear negative association between the caste status of the respondents and their mean number of live-births in Ghudani Kalan, and with one exception, in Ghagga as also for all the three villages. But there was no clear trend in Ugala. In this village the highest number of live-births were observed among middle castes and as their number was quite small, therefore, no generalization could be made for them. It was further noticed that the second highest number of live-births were found among scheduled caste respondents and the high castes as well as the backward caste respondents had the same number of live-births.

Another important point noted was that given the caste categories, there were significant variations in the number of

TABLE 7.1

Caste status and mean number of live-births

Caste status	Ugala		Ghagga		Ghudani Kalan		Total	
	Mean number of live-births	Number and % of respondents	Mean number of live-births	Number and % of respondents	Mean number of live-births	Number and % of respondents	Mean number of live-births	Number and % of respondents
Higher Castes	4.55	99 (42.12)	4.65	73 (26.74)	3.78	26 (8.70)	4.48	198 (24.50)
Middle Castes	6.28	11 (4.68)	4.14	58 (21.25)	3.86	188 (62.67)	4.03	257 (31.81)
Backward Castes	4.55	60 (25.53)	4.91	33 (12.09)	4.30	37 (12.37)	4.67	130 (16.09)
Scheduled Castes	5.23	65 (27.66)	5.04	109 (39.92)	5.20	49 (16.39)	5.13	223 (27.60)
Total	4.82	235 (100.)	4.73	273 (100.)	4.12	300 (100.)	4.53	808 (100.)

live-births in the three villages in the first two caste categories, i.e., among upper and middle caste respondents, but these variations were less marked in the last two caste categories, indicating thereby that backward and scheduled castes, on the whole, had more or less the same level of fertility in all these villages.

Thus, on the whole, in the total sample as well as in each village separately with one exception, scheduled caste respondents had comparatively higher number of live-births and this caste was followed by other backward castes. Moreover, the association between the caste status and mean number of live-births was clear in Ghagga and Ghudani Kalan but no clear trend was observed in Ugala.

Since there was no consistent trend in the relationship between caste status and number of live-births in all these villages an attempt was made to examine this relationship within a given duration of married life as more significant relationship could be observed by controlling the duration of married life. Table 7.2 gives mean number of live-births by the duration of marriage in each caste category.

In Table 7.2 there was no clear trend in the shorter duration, i.e., less than 20 years of married life, although, even during this duration, mean number of live-births was comparatively higher among scheduled castes as compared with the other castes. But in the longer duration of married life, i.e., 20 years or more, where the respondents had either completed their reproductive period, or they were near completion, mean number of live-births was much higher among scheduled caste respondents for the total sample as well as for all the villages. It was further noted that given the same caste categories, with few exceptions where the number of respondents was small, mean number of live-births was higher in Ugala and Ghagga and little lower in Ghudani Kalan. Thus, in the longer duration of marriage, i.e., 20 years and more, fertility differentials were quite significant between the upper and the lower castes.

Since it is assumed that caste status affects fertility indirectly through other intervening variables, an attempt has been made to examine the role of age at marriage and the percentage of child loss on fertility in each caste category.

TABLE 7.2

Mean number of live-births and duration of marriage by caste status

Caste status	Ugala		Ghagga		Ghudani Kalan		Total	
	Mean number of live-births	Number and % of respondents	Mean number of live-births	Number and % of respondents	Mean number of live-births	Number and % of respondents	Mean number of live-births	Number and % of respondents
Upper Caste	2.84	6.33	2.63	5.81	2.85	4.67	2.77	5.89
Middle caste	3.67	7.12	2.84	5.52	2.46	5.04	2.58	5.26
Backward caste	2.76	6.05	2.41	9.00	2.91	5.30	2.68	6.30
Scheduled caste	3.60	7.18	2.79	9.00	3.55	6.92	3.18	7.90

Therefore, Table 7.3 gives mean number of live births in each caste category according to the age at marriage and percentage of child deaths.

TABLE 7.3

Mean number of live-births, age at marriage and percentage of children dead by caste status

Caste status	Mean age at marriage	Percentage of child deaths	Mean number of live-births
Ugala			
Upper caste	18.18	11.97	4.56
Middle caste	16.33	4.29	6.28
Backward caste	17.60	11.31	4.55
Scheduled caste	16.18	16.76	5.23
Ghagga			
Upper caste	18.07	10.00	4.65
Middle caste	18.17	12.92	4.14
Backward caste	17.59	15.34	4.91
Scheduled caste	17.44	16.61	5.31
Ghudani Kalan			
Upper caste	18.27	10.10	3.78
Middle caste	19.95	16.94	3.86
Backward caste	17.09	16.99	4.30
Scheduled caste	16.56	25.88	5.20
Total Sample			
Upper caste	18.15	11.01	4.48
Middle caste	17.93	15.51	4.03
Backward caste	17.45	15.93	4.67
Scheduled caste	16.88	18.72	5.12

It is clear from the table that in all the three villages individually as also in the total sample, scheduled caste respondents had lower rather the lowest age at marriage, and next, in order with one exception, were respondents belonging to backward castes. On the other hand respondents belonging to higher castes had comparatively higher age at marriage. It was interesting to note that in Ugala where respondents belonging to the middle castes had the highest number of live-births their age at marriage was also comparatively lower though slightly higher than the scheduled castes. Thus, mean age at marriage was comparatively lower among scheduled caste and backward caste respondents. Therefore, their higher number of live-births could partly be attributed to their lower age at marriage.

The extent of infant and child loss is another important factor that influences the fertility level. In Table 7.3 the extent of child loss or the percentage of dead children varies inversely with the caste status of the respondents for the total sample as well as for Ghudani Kalan and Ghagga. Since the percentage of dead children was the highest in the total sample as well as in the above mentioned villages, their higher number of live-births could also be due to this reason. In Ugala there was no clear trend in the percentage of child deaths, consequently, there was no clear trend in the mean number of live-births also.

Thus, on the whole, comparatively lower age at marriage and higher percentage of child deaths, to some extent, might have been responsible for higher number of births among scheduled caste respondents.

Since socio-economic status is also an important determinant of fertility behaviour, therefore, an attempt has also been made to examine the role of socio-economic factors as intervening variables in each caste category to determine its influence on fertility level, their husband's occupational status, average family income and their level of living in each caste category.

In Table 7.4, educational status of the respondents varies inversly with their caste status, and the percentage of illiterate respondents was the highest, i.e., nearly 95 per cent among scheduled caste respondents in all these villages. Further, even the educated among the scheduled caste respondents, none had education beyond matriculation level. Thus, educationally they were at a very low level as compared with other caste groups.

Even the husband's occupational status, varied inversely with the caste status of the respondents and large majority of the

TABLE 7.4(a)

Distribution of respondents by their level education, occupational status and average family income in each caste of category in Ugala and Ghagga

Educational/ occupational/ income level	Ugala				Ghagga			
	Upper caste No. & % of respondents	Middle caste No. & % of respondents	Bacakward caste No. & % of respondents	Scheduled caste No. & % of respondents	Upper caste No. & % of respondents	Middle caste No. & % of respondents	Backward caste No. & % of respondents	Scheduled caste No. & % of respondents
Educational level								
Illiterate	59 (59.60)	9 (81.82)	47 (78.33)	62 (95.38)	53 (72.60)	48 (82.76)	30 (90.91)	109 (100.00)
Below matric	29 (29.29)	1 (9.09)	11 (18.33)	3 (4.62)	14 (19.18)	7 (12.07)	3 (9.09)	—
Matriculation and above	11 (11.11)	1 (9.09)	2 (3.34)	—	6 (8.22)	3 (5.17)	—	—
Total	99 (100.00)	11 (100.00)	60 (100.00)	65 (100.00)	73 (100.00)	58 (100.00)	33 (100.00)	109 (100.00)

Occupational prestige categories								
1—3	20 (20.20)	1 (9.09)	2 (3.33)	—	12 (16.44)	2 (3.45)	2 (6.06)	—
4—5	67 (67.68)	7 (63.64)	36 (60.00)	9 (13.85)	46 (63.01)	50 (86.21)	21 (63.64)	22 (20.18)
6—7	7 (7.07)	2 (18.18)	19 (31.67)	54 (85.08)	11 (15.07)	1 (1.72)	9 (27.27)	72 (72.47)
No occupation	5 (5.05)	1 (9.09)	3 (5.00)	2 (3.07)	4 (5.48)	5 (8.62)	1 (3.03)	8 (7.34)
Total	99 (100.00)	11 (100.00)	60 (100.00)	65 (100.00)	73 (100.00)	58 (100.00)	33 (100.00)	109 (100.00)
Income								
Average yearly income	Rs. 13,737	Rs. 9,886	Rs. 9,604	Rs. 3,463	Rs. 8,332	Rs. 11,961	Rs. 7,917	Rs. 4,037

TABLE 7.4(b)

Distribution of respondents by their level of education, occupational status and average family income in each caste category in Ghudani Kalan and Total Sample (Ugala, Ghagga and Ghudani Kalan)

Educational/ occupational/ income level	Ghudani Kalan				Total Sample			
	Upper caste	Middle caste	Backward caste	Scheduled caste	Upper caste	Middle caste	Backward caste	Scheduled caste
	No. & % of respondents	No. & % of respondents	No. & % of respondents	No. & % of respondents	No. & % of respondents	No. & % of respondents	No. & % of respondents	No. & % of respondents
Educational Level								
Illiterate	13 (50.00)	130 (69.15)	31 (83.78)	45 (51.84)	125 (63.13)	187 (72.76)	108 (83.08)	216 (96.86)
Below matric	8 (30.77)	40 (21.28)	5 (13.52)	4 (8.16)	51 (25.76)	48 (18.68)	19 (14.62)	7 (13.14)
Matriculation and above	5 (19.23)	18 (19.57)	1 (2.70)	—	22 (11.11)	22 (8.56)	3 (2.30)	—
Total	26 (100.00)	188 (100.00)	33 (100.00)	49 (100.00)	198 (100.00)	257 (100.00)	130 (100.00)	223 (100.00)

Occupational prestige categories								
1—3	3 (11.54)	2 (1.06)	1 (3.85)	—	35 (17.58)	5 (1.95)	5 (3.84)	—
4—5	13 (50.00)	135 (71.81)	13 (53.85)	6 (16.59)	125 (63.63)	192 (74.71)	70 (53.85)	37 (16.59)
6—7	—	10 (5.32)	16 (33.84)	38 (76.68)	18 (9.39)	13 (5.06)	44 (33.85)	171 (76.68)
No occupation	10 (38.46)	41 (21.81)	— (8.46)	5 (6.73)	19 (9.60)	47 (18.28)	11 (8.46)	18 (6.43)
Total	26 (100.00)	188 (100.00)	37 (100.00)	49 (100.00)	191 (100.00)	157 (100.00)	130 (100.00)	223 (100.00)
Income								
Average yearly income	Rs. 10,913	Rs. 13,610	Rs. 3,784	Rs. 4,184	Rs. 11,370	Rs. 13.130	Rs. 7,519	Rs. 3,901

respondents from the upper and the middle castes had their husbands engaged in higher prestige occupations, whereas, large majority of the scheduled and other backward caste respondents had their husbands engaged in lower occupations which carry low prestige and have low income level in all these villages.

Similar differences were observed in the average annual family income, being the lowest among the scheduled caste respondents and the highest among the upper caste respondents.

Thus, on the whole, the data clearly reveal that the scheduled caste respondents were at the lowest level in the terms of their socio-economic conditions and the differences between the upper castes and lower castes were quite significant. Nevertheless, the distinctions between the upper caste and middle caste respondents in terms of their social and economic conditions were not very sharp.

Another index of the socio-economic status is the level of living of the family. Table 7.5 also shows that in the terms of level of living, large majority of the upper and middle caste respondents fell in the category of higher level of living whereas reverse was true in case of backward and scheduled caste respondents.

TABLE 7.5(a)

Caste status and level of living

Ugala

Caste status	Low level No. & % of respondents	Medium level No. & % of respondents	High level No. & % of respondents	Total
Upper caste	14 (14.14)	21 (21.21)	64 (64.65)	99 (100.00)
Middle caste	3 (27.27)	2 (18.18)	6 (54.55)	11 (100.00)
Backward caste	20 (33.33)	20 (33.33)	20 (33.33)	60 (100.00)
Scheduled caste	53 (81.54)	9 (13.85)	3 (4.61)	65 (100.00)
Total	90 (38.30)	52 (22.13)	93 (39.57)	235 (100.00)

TABLE 7.5(b)

Caste status and level of living

Ghagga

Caste status	Low level No. & % of respondents	Medium level No. & % of respondents	High level No. & % of respondents	Total
Upper caste	18 (24.66)	15 (20.55)	40 (54.79)	73 (100.00)
Middle caste	10 (17.24)	17 (19.31)	31 (53.45)	58 (100.00)
Backward caste	5 (15.15)	13 (39.39)	15 (45.46)	33 (100.00)
Scheduled caste	48 (44.04)	47 (43.12)	15 (12.84)	109 (100.00)
Total	81 (29.67)	92 (33.70)	100 (36.63)	273 (100.00)

TABLE 7.5(c)

Caste status and level of living

Ghudani Kalan

Caste status	Low level No. & % of respondents	Medium level No. & % of respondents	High level No. & % of respondents	Total
Upper caste	1 (3.85)	5 (19.23)	20 (76.92)	26 (100.00)
Middle caste	13 (6.91)	25 (13.30)	150 (79.79)	188 (100.00)
Backward caste	14 (37.84)	15 (40.54)	8 (21.62)	37 (100.00)
Scheduled caste	22 (44.90)	18 (36.73)	9 (18.37)	49 (100.00)
Total	50 (16.67)	63 (21.00)	187 (62.33)	300 (100.00)

TABLE 7.5(d)

Caste status and level of living

Total Sample

Caste status	Low level No. & % of respondents	Medium level No. & % of respondents	High level No. & % of respondents	Total
Upper caste	33 (16.67)	41 (20.71)	124 (62.63)	198 (100.00)
Middle caste	26 (10.12)	44 (17.12)	187 (72.76)	257 (100.00)
Backward caste	39 (30.00)	48 (36.92)	43 (33.08)	130 (100.00)
Scheduled caste	123 (55.16)	74 (33.18)	26 (11.66)	223 (100.00)
Total	221 (27.35)	207 (25.62)	380 (47.03)	808 (100.00)

It may, thus, be concluded that mean number of live-births were comparatively higher among scheduled caste respondents as they were educationally, economically and socially backward. Their age at marriage was also low. Their living conditions and their environment are generally responsible among them for the higher incidence of child loss which in turn results in their higher number of births.

Use of contraception is another important variable that influences fertility behaviour although its influence on fertility may be either direct or indirect through socio-economic status. But here it is expected that the use of contraception would be more prevalent among upper and middle caste respondents due to their higher educational level and better economic status. Table 7.6 gives the use of contraception by the caste status of the respondents.

Contrary to our expectations the use of contraception did not show any correlation with the caste status of the respondents as there was no consistent trend in the total sample as well as in each village separately.

TABLE 7.6

Use of contraception by caste status

Caste status	Ugala: No. & % of users	Ugala: No. & % of Non-users	Ugala: Total No. & % of respondents	Ghagga: No. & % of users	Ghagga: No. & % of Non-users	Ghagga: Total No. & % of respondents
Higher caste	35 (35.35)	64 (64.65)	99 (42.13)	26 (35.62)	47 (64.39)	73 (26.74)
Middle caste	2 (18.18)	9 (81.82)	11 (4.68)	27 (46.55)	31 (53.45)	58 (21.25)
Backward caste	27 (45.00)	33 (55.00)	60 (25.83)	13 (39.39)	20 (60.61)	33 (12.08)
Scheduled caste	21 (32.31)	44 (67.69)	65 (27.66)	34 (31.19)	75 (68.81)	109 (39.93)
Total	85 (36.17)	150 (63.83)	235 (100.)	100 (36.63)	173 (63.37)	273 (100.)

Caste status	Ghudani Kalan			Grand Total		
	No. & % of users	No. & % of Non-users	Total No. & % of respondents	No. % & of users	No. & % of Non-users	Total No. & % of respondents
Higher caste	6 (23.08)	20 (76.20)	26 (8.67)	67 (33.84)	131 (66.17)	198 (24.50)
Middle caste	47 (25.00)	141 (75.00)	188 (62.67)	76 (29.57)	181 (70.43)	257 (31.81)
Backward caste	8 (21.62)	29 (78.38)	37 (12.33)	48 (36.92)	82 (63.08)	130 (16.09)
Scheduled caste	16 (32.65)	33 (67.35)	49 (16.33)	71 (31.84)	152 (68.17)	223 (27.60)
Total	77 (25.67)	223 (74.33)	300 (100.)	262 (32.43)	546 (67.57)	808 (100.)

Lastly, the conception of a desired ideal family size could be another important variable which can influence actual family size although the desired ideal family size need not necessarily be the actual family size. It only indicates what one thinks should be the size of an ideal family.

The analysis of the data did not show any clear association between the caste status of the respondents and the number of children specified for an ideal family in the total as well as in every village, therefore, the same is not presented in a table.

To summarize, it may be stated that there was close association between caste status and mean number of live-births in the total sample as well as in Ghudani Kalan and Ghagga, especially, when controlled for marriage duration but no clear trend was observed in Ugala. The data also supported the assumption that it is not caste per se which influence fertility as the scheduled caste and also backward caste respondents, on the whole, had low socio-economic level, lower age at marriage and higher incidence of child deaths, therefore, among them higher number of births could be due to these reasons.

REFERENCES

1. Agarwala, S.N. 1977: India's Population Problems, New Delhi, Tata McGraw Hill Publishing Company.
2. Driver, E.D. 1960: "Fertility Differentials among Economic Strata in Central India", Eugenic Quarterly, 7, 2, June.
3. Wyon, John B. and Gordon, John E. 1971: The Khanna Study, Cambridge, Harvard University Press
4. Rele, J.R. 1963: "Fertility Differentials in India—Evidence from a Rural Background", Milbank Memorial Fund Quarterly, Vol. XII, 2, April.
5. Sexena, G.B. 1965: Differential Fertility in Rural Hindu Community: A Sample Survey of Rural Uttar Pradesh, India, Eugenic Quarterly, 12, (3) Sept.
6. United Nations, 1961: The Mysore Population Study, New York.

8

RELIGION AND FERTILITY

RELIGION is considered to be one of the important variables that determines fertility behaviour. It is a major source of values which govern the normative behaviour in various sphears, including, fertility through religious sanctions. As stated by Chaudhury (1982), religion prescribes a code of life, refers to a system of beliefs, attitudes and practices which individuals share in groups and through this orientation towards life and deaths, it is supposed to affect one's fertility behaviour.

Differentials in fertility due to religious and other cultural practices have been reported in many countries and in India, too, religion is an important cause of fertility differentials.

A number of studies conducted at the national level, as well as in different parts of the country, bring out the religious differences in fertility behaviour of major religious groups—the Hindus and the Muslims. The decadal growth rate of population for Hindus, Muslims, Christians, Sikhs, Buddhists and Jains was reported to be 23.69, 30.84, 32.58, 32.28, 17.33 and 28.49 respectively in the intercensal period—1961-71 (Census 1971). Among all the religious groups, the highest rate of population growth is observed among Christians during the decade 1961-71. But, according to the results of a fertility survey in a sub sample of Sample Registration scheme, 1972, the Muslims have the highest fertility, whereas, it is the lowest for the Christians and the Hindus fall in between, both in urban and rural parts of the country. Mysore Study (UN 1961) also indicates that religion is associated with fertility; married muslim women have borne, on an average, larger number of children than Hindu women. Rele and Kanitkar (1976) in their study in Greater Bombay, have estimated, after controlling the educational attainments and age, the differentials between the Hindus and the Sikhs, and between the Hindus and others, narrowed down but did not disappear. Standardised average number of living children among Hindus, Muslims and Christians was found to be 2.65, 2.95 and 2.80, respectively.

There are two basic assumptions for analysing the religious differentials in fertility. According to one, the impact of religion on fertility related attitudes and behaviour operates with particular religious doctrine only, affecting its social norms and religion affects fertility through these norms. For example, among the Hindus the influence of religion on fertility can be seen through son preference as sons have an important role in religious ceremonies and for the performance of the last rites of a man. The second assumption suggests that the particular fertility levels of a religious group indicate only the socio-economic characteristics of that group. Whatever may be the assumptions the fact remains that religion is an important cause of fertility differentials.

Therefore, the present chapter is an attempt to examine fertility differentials among different religious groups in the sample population. An attempt would also be made to analyse these differentials by the level of religiosity of the respondents. It is assumed that religion per se may not be an important determinant of fertility in the present study because the two major religious groups are the Hindus and the Sikhs, and the number of the Muslims is quite small, and the normative practices between these two major religious groups are not very dissimilar. Whatever differentials are there, these might be due to the variations in the demographic and socio-economic characteristics of these respondents. Table 8.1 gives mean number of live-births by the religion of the respondents.

The data reveal that Ugala and Ghagga had larger number of Hindu respondents, whereas, Ghudani Kalan had larger number of Sikh respondents. However, it is interesting to observe that the mean number of live-births was comparatively higher among the Hindu respondents in the two villages dominated by the Hindu respondents and reverse was true in Ghudani Kalan which had larger number of Sikh respondents although the difference between the two religious communities was small. Compared to these two communities, the Muslim

TABLE 8.1

Religion Vs. mean number of live-births

Religious categories	Ugala		Ghagga		Ghudani Kalan		Total	
	Mean number of live-births	Number & % of respondents	Mean number of live-births	Number & % of respondents	Mean number of live-births	Number & % of respondents	Mean number of live-births	Number & % of respondents
Hindu	4.88	189 (80.43)	4.85	178 (65.20)	3.95	42 (14.00)	4.77	409 (50.62)
Muslim	—	—	4.25	4 (1.47)	3.85	7 (2.33)	4.00	11 (1.36)
Sikh	4.62	46 (19.57)	4.51	91 (33.33)	4.15	251 (83.67)	4.29	388 (48.02)
Total	4.82	235 (100.)	4.73	273 (100.)	4.12	300 (100.)	4.53	808 (100.)

respondents had smaller number of live-births but their number was too small to justify making any valid generalizations. The data also reveal some inter-village variations in the mean number of live-births within the same religious categories.

Since the variations in the mean number of live-births were not very striking, an analysis of the specific duration of marriage within these religious categories might give us better idea of these differentials. Therefore, Table 8.2 gives mean number of live-births among these religious categories by marriage duration.

In Table 8.2 similar trend was observed in the mean number of live-births for all the villages. The significant point noted was that in the longer duration of marriage, i.e., 20 years and more, the variations in the mean number of live-births were quite significant (more than 1 child in the total sample as well as for Ugala and Ghagga individually). But the difference was almost negligible in Ghudani Kalan indicating thereby that in this village although the Sikh respondents had higher number of live-births but it was not significantly higher than that of the Hindus. Another notable fact was that in the longer marriage duration the Muslim respondents, though small in number, had the highest number of live-births.

As pointed out earlier, since religion per se is not considered to be an important determinant of fertility behaviour, in these two major religious communities because of similarity in their normative practices, an attempt has been made to analyze the role of age at marriage, the extent of child loss and use of contraception among these respondents as intermediate variable through which religion might have influenced fertility behaviour. Table 8.3 gives distribution of respondents by their age at marriage, percentage of child deaths and use of contraception.

So far as age at marriage was concerned, the data clearly revealed comparatively higher age at marriage for the Sikh respondents as against the Hindus in each village, and the differences in the age at marriage were well marked in Ugala and Ghagga where the variations in the mean number of live-births were also greater, whereas, the difference was less in Ghudani Kalan and the variation in the mean number of live-births was also low. Thus, on the whole, it indicates that in Ugala and

TABLE 8.2

Mean number of live-births and duration of marriage by religion

Religious categories	Ugala		Ghagga		Ghudani Kalan		Total	
	Marriage duration		Marriage duration		Marriage duration		Marriage duration	
	0-19 Yrs Mean number of live-births	20+Yrs Mean number of live-births	0-19 Yrs Mean number of live-births	20+Yrs Mean number of live-births	0-19 Yrs Mean number of live-births	20+Yrs Mean number of live-births	0-19 Yrs Mean number of live-births	20+Yrs Mean number of live-births
Hindu	3.10	6.90	2.63	7.39	2.71	5.26	2.84	6.92
Muslim	—	—	3.00	8.00	3.02	7.00	3.01	7.33
Sikh	3.00	5.25	2.85	6.32	2.75	5.30	2.79	5.50

TABLE 8.3

Mean number of live-births, age at marriage, percentage of child deaths and use of contraception by religious categories

	Ugala			Ghagga			Ghudani Kalan			Total sample		
	Hindu	Muslim	Sikh	Hindu	Muslim	Sikh	Hindu	Muslim	Sikh	Hindu	Muslim	Sikh
Mean no. of live-births	4.88	—	4.62	4.85	4.25	4.51	3.95	3.85	4.15	4.77	4.00	4.29
Mean age at marriage in yrs.	17.51	—	18.50	17.81	16.50	18.07	17.83	15.83	17.89	17.67	16.08	18.00
Percentage of children dead	13.56	—	9.39	14.37	—	13.87	13.86	11.11	19.19	13.93	6.82	16.63
Users of contraception (Percentage)	32.80	—	50.00	34.27	50.00	40.66	28.57	14.29	25.50	33.01	27.27	31.96
Non-users of contraception (Percentage)	67.20	—	50.00	65.73	50.00	59.34	71.43	85.71	74.50	66.99	72.73	68.04

Ghagga, the Hindu respondents had higher number of live-births but their age at marriage was comparatively lower than that of the Sikh respondents although the same was not true for Ghudani Kalan.

A glance at the percentage of child deaths in the two major religious categories again indicates similar trend. It was a matter of strange coincidence that the percentage of child deaths was also higher among the Hindu respondents in Ugala and Ghagga where they had higher number of live-births and reverse was true for the Sikh respondents where they had higher percentage of child deaths.

The data about the use of contraception again indicate similar trend as the use of contraception was comparatively more prevalent among the Sikh respondents in Ugala and Ghagga and opposite was true for Ghudani Kalan and the difference between the two categories was also small for this village.

Thus, on the whole, the data indicate that the Hindu respondents in Ugala and Ghagga had higher number of live-births, at the same time they also had comparatively lower age at marriage, higher incidence of child deaths and lesser use of contraception and opposite was true for the Sikh respondents in Ghudani Kalan.

In terms of socio-economic differences no significant variations were observed among these two religious categories except in the level of living which was found to be higher among the Sikh respondents in all these villages, therefore, fertility differentials in these categories were not analysed by their socio-economic status.

Thus, in the absence of any significant variations in the socio-economic status of these two religious categories, the variations in the mean number of live-births could be due to their lower age at marriage, higher incidence of child mortality and lesser use of contraception and may not be due to religion per se.

Religiosity and Fertility

Although religion is an important determinant of fertility but one's mere faith in religion may not truly indicate the impact of religion on fertility because mere identification with a religious faith may not necessarily reflect one's commitment to that religion. This can be understood better by knowing one's attitudes towards religion and religious activities. In other words, it becomes more important to know how strongly a person feels about his religion.

There are hardly any studies in India which deal with the relationship between the extent of religiosity and fertility, although scholars in other countries have tried to understand this relationship. For example, Chaudhury (1980) found a strong inverse relationship between religiosity and use of contraception and a positive relationship between religiosity and fertility among a cross section of women in Dacca city. This relationship holds true even after allowance is made for the effect of other socio-economic variables. He concludes that religiosity has an independent effect on fertility behaviour over and above one's socio-economic status. Westoff et al. (1963) showed that the Roman Catholics in U.S.A. were influenced by their religious belief in their fertility behaviour. They found that active (those who attend church regularly also abide by other religious prescriptions) Catholics generally had higher actual and desired fertility in comparison with non-active Catholics and other religious groups.

It, therefore, appears that the higher the degree of commitment to the religious faith, the higher would be the fertility level and lesser use of family planning methods. As already stated, in the present study also an index of religiosity was prepared by taking into account some of the statements regarding the performance of certain religious activities of the respondents. Accordingly, as stated earlier, the respondents were divided into three categories, namely, highly religious, moderately religious and less or least religious. Table 8.4 gives mean number of live-births by the level of religiosity of the respondents.

Table 8.4 indicates that although mean number of live-births was comparatively higher among highly religious respondents in Ghagga and Ghudani Kalan but reverse was not true for less or least religious respondents. In Ugala also there was no clear trend. Moreover, in all the villages moderately religious

TABLE 8.4

Level of religiosity and mean number of live-births

Level of religiosity	Ugala		Ghagga		Ghudani Kalan		Total	
	Mean number of live-births	Number & % of respondents	Mean number of live-births	Number & % of respondents	Mean number of live-births	Number & % of respondents	Mean number of live-births	Number & % of respondents
High level of religiosity	4.84	200 (85.11)	5.05	188 (68.86)	4.32	237 (79.00)	4.71	6.25 (77.35)
Medium level	4.77	26 (11.06)	3 81	54 (19.78)	3.33	42 (14.00)	3.85	122 (15.10)
Low level	5.11	9 (3.83)	4.30	31 (11.36)	3.42	21 (7.00)	4.11	61 (7.55)
Total	4.82	235 (100.)	4.72	273 (100.)	4.12	300 (100.)	4.53	808 (100.)

respondents had lesser number of live-births as compared to the other two categories.

Even when controlled for the duration of marriage, no clear trend was observed, indicating thereby lack of any association between the two. Since the number of respondents in the category of less religious respondents was quite small, therefore, no generalization could be made.

In the absence of any clear trend, an analysis of the other variables such as the extent of child mortality, use of contraception and desired family size also did not show any association between the level of religiosity and mean number of live-births.

It may, thus, be stated that, on the whole, the Hindu respondents had higher number of live-births than the Sikh respondents in the total sample as well as in Ugala and Ghagga, but reverse was true for Ghudani Kalan. It was, further, observed that the Hindu respondents has comparatively lower age at marriage, higher incidence of child mortality and lesser use of contraception as compared with the Sikh respondents. However, the level of religiosity did not show any association with the number of live-births in any village.

REFERENCES

1. Census Commissioner, India 1971: Census of India, 1971, Social and Cultural tables, Series I, Part II-C (i), Office of the Registrar General.
2. Census Commissioner, India 1972: Census of India, 1971, Social and Cultural tables, Series I, Part II-C (i), Office of the Registrar General.
3. Chaudhury, R.H. 1982: Social Aspects of Fertility, New Delhi, Vikas Publishing House.
4. Rele, J.R. and Kanitkar 1977: "Fertility Differentials by Religion in Greater Bombay: Role of Explanatory Variables" in the Economic and Social Supports for High Fertility (ed.) L.T. Ruzicka, Canberra, Australian National University.
5. United Nations 1961: Mysore Population Study, New York.
6. Vasaria, Lella: "Religious Differentials in Fertility in Population in India's Development from 1947 to 2000" (ed.) Ashish Bose, Mitra, New Delhi, Vikas Publishing House.
7. Wastoff, C.F., et al. 1961: The Third Child: A Study in the Prediction of Fertility, New Jersey, Princetion University Press.

9

MODERN AND TRADITIONAL VALUE ORIENTATIONS AND FERTILITY

In the present chapter an attempt has been made to examine the role of modern and traditional value orientations of the respondents in their fertility behaviour. As already mentioned, large number of studies have highlighted the importance of socio-economic factors as the major determinants of fertility behaviour but the basic question as suggested by Clifford (1971) is what leads members of one class or group to prefer small families and to practice limitation while members of another class or group do not. Kiser and Whelpton also (1953) believed that it is not the socio-economic status per se but rather some other underlying factor(s) characteristic of these classes that account for the variations in fertility behaviour. In attempting to determine what influences these differences in fertility behaviour, Rainwater (1960) suggests that we turn to the "goals and orientations to the world", which couples reveal in their attitudes towards family planning. Clifford (1971) maintained that the complex of socio-economic forces which has shown marked effects on fertility exert their influence through the determination and reinforcement of certain value orientations. Mukherjee (1979), while analysing the modernity orientations of Indian women, maintains that modernity values are a coherent psychological syndrome that results from particular processes of socialization, which occurs only in particular social settings. According to him, urban residence, formal schooling, higher socio-economic status, greater exposure to mass media, etc., facilitate the acquisition of modernity values, which, in turn, have a dampening effect on fertility.

Empirical evidence collected in India regarding the relationship between modernity value orientations and fertility is not very substantial, although Pareek and Kothandapani (1969) reported a significant correlation between ideal family size and birth control, on the one hand, and some indices of individual modernity, on the other. For their study, they analysed the Harvard Project data collected in the tribal belt of Bihar, in North-Eastern India, and found that education, personal modernity, political modernization, overall modernization and lack of fatalism accounted for 10 per cent of the variance in preference for a small family. George (1972) reported a direct relationship between traditionalism and fertility. Mukherjee's (1979) study revealed mixed findings. The correlational analysis revealed that knowledge of contraception and attitude towards family planning were almost consistently related in a positive direction to all the three of the modernity values but they showed a weak relationship with contraceptive adoption.

The present chapter as stated earlier in an attempt to examine fertility differentials among these respondents by their modernity value orientations. It is assumed that the higher the level of modernity, the greater the adoption of family planning practices and the smaller the actual and desired family size.

As stated in the methodology, an index of the level of modernity orientations of the respondents was prepared by taking into considerations some statements reflecting their proneness to change The responses given by each respondents were scored, and, accordingly, the respondents were divided into three categories, namely, traditional, moderately modern and modern. Thus, according to the total score, 39 per cent of the respondents came in the category of traditional, 41 per cent as moderately modern and the remaining 20 per cent were modern.

Table 9.1 gives the distribution of respondents according to their modernity value orientations and mean number of live-births.

The data in Table 9.1 reveal close association between the level of modernity and mean number of live-births in the total sample as well as in Ghagga and Ghudani Kalan. In Ugala although mean number of live-births were found to be the highest among traditional respondents, modern respondents had slightly higher number of live-births as compared to the moderately modern. It was further noted that the difference

TABLE 9.1

Distribution of respondents by their mean number of live-births and modernity value orientation

Level of modernity	Ugala		Ghagga		Ghudani Kalan		Total	
	Mean no. of live-births	No. & % of respondents	Mean no. of live-births	No. & % of respondents	Mean no. of live-births	No. & % of respondents	Mean no. of live-births	No. & % of respondents
Traditional	5.19	120 (51.06)	5.39	91 (33.33)	4.97	107 (35.67)	5.17	318 (39.36)
Moderately modern	4.32	86 (36.60)	4.51	119 (43.59)	3.85	124 (41.33)	4.21	329 (40.72)
Modern	4.83	29 (12.34)	4.20	63 (23.08)	3.30	69 (23.00)	3.93	161 (19.92)
Total	4.82	235 (100.)	4.73	273 (100.)	4.12	300 (100.)	4.53	808 (100.)

in the mean number of live-births between traditional and modern respondents was quite significant.

Although, on the whole, the relationship in Table 9.1 is quite clear, still, in order to eliminate the influence of any chance factor, an attempt was made to examine this relationship by controlling the duration of marriage. But when duration of marriage was controlled a similar trend was observed confirming the fact that the level of modernity was negatively associated with the number of live-births.

Since it is not modernity values per se which depress fertility but these value orientations operate through certain psychological processes in given social settings, it is proposed to examine the role of age at marriage, percentage of child deaths, level of interspouse communication and the use of contraception in causing these differentials. Table 9.2 gives the distribution of respondents according to their level of modernity, their age at marriage, extent of child deaths, interspouse communication and use of contraception.

So far as age at marriage is concerned the data clearly reveal higher age at marriage for modern respondents as against traditional one's in every village, although the difference was not very substantial.

Similar trend was observed in case of percentage of child deaths as in every village modern respondents had comparatively lower percentage of child deaths.

So far as interspouse communication regarding the use of contraception was concerned, again the percentage of communicators was comparatively found to be higher among modern respondents although there were some discrepancies between the other two categories.

The use of contraception is an important variable through which modern values influence fertility behaviour. Here, too, the data reveal comparatively higher proportion of users among modern respondents for the total sample as well as for every village separately.

Thus, on the whole modern respondents had lesser number of live-births as compared with traditional respondents, but, at the same time, they also had higher age at marriage, lower incidence of infant and child deaths, interspouse communication

TABLE 9.2

Distribution of respondents by their modernity value orientation, age at marriage, extent of child deaths, level of interspouse communication and use of contraception

	Ugala			Ghagga		
	Traditional	Moderately modern	Modern	Traditional	Moderately modern	Modern
Age at marriage in yrs.	17.30	17.88	18.78	17.92	17.67	18.23
Child mortality	(3.96) 87	(11.86) 44	(10.14) 14	(14.90) 73	(13.06) 70	(14.37) 38
Interspouse communication						
Communicators	37 (30.83)	18 (20.93)	8 (27.59)	19 (20.88)	16 (13.44)	19 (30.16)
Non-communicators	83 (69.17)	68 (79.07)	21 (72.41)	72 (79.12)	103 (86.56)	44 (69.84)
Total	120	86	29	91	119	63
Use of contraception						
Users	43 (35.83)	29 (33.72)	13 (44.83)	28 (30.77)	48 (38.66)	26 (41.27)
Non-users	77 (64.17)	57 (66.28)	16 (55.17)	63 (69.23)	73 (61.34)	37 (58.73)

	Ghudani Kalan			Total		
	Traditional	Moderately modern	Modern	Traditional	Moderately modern	Modern
Age at marriage in yrs.	16.69	18.15	18.95	17.27	17.91	18.64
Child mortality	(18.42) 98	(20.55) 98	(13.22) 30	(15.68) 258	(15.32) 212	(13.02) 82
Interspouse communication						
Communicators	13 (12.15)	31 (25.00)	20 (28.98)	69 (21.70)	65 (19.75)	47 (35.49)
Non-communicators	94 (87.85)	93 (75.00)	49 (71.02)	249 (78.30)	264 (80.25)	114 (70.81)
Total	107	124	69	318	329	161
Use of contraception						
Users	23 (21.50)	32 (25.81)	22 (31.88)	94 (29.56)	107 (32.52)	61 (37.89)
Non-users	84 (78.50)	92 (74.19)	47 (68.12)	224 (70.44)	222 (67.48)	100 (62.11)

(Percentages are given in parenthesis.)

and greater use of contraception and these might have contributed to the smaller number of live-births among modern respondents.

Another important variable which influences fertility behaviour is the conception of desired ideal family size although desired ideal size is that number which the respondents would like to have and not what they actually have. It has often been noted that the preference for an ideal family size is influenced by the actual number of children in the family. Various studies have revealed that most often whatever number of children women have, they consider that to be the ideal family size. Therefore, Table 9.3 gives the desired ideal family size and actual family size by the level of modernity.

Table 9.3, again, reveals similar trends as it indicates negative relationship between the level of modernity and the desired ideal family size, on the one hand and the actual number of live-births, on the other, for the total sample as well as for Ghagga and Ghudani Kalan. Only in Ugala no clear trend emerged.

Lastly, socio-economic status of the respondents did not reveal any significant variation from the modernity value orientations of the respondents, therefore, such an analysis was not attempted.

It may thus be summarised that, on the whole, modernity value orientations of the respondents did reveal negative association with the number of live-births for the total sample as well as for Ghagga and Ghudani Kalan and some of the variables that accounted for these variations were age at marriage, interspouse communication and use of contraception.

REFERENCES

1. Clifford, William B. 1971: "Modern and Traditional Value Orientations and Fertility Behaviour: A Social Demographic Study" in Demography, Vol. 18, 1, Feb., 1971.
2. Kiser, Clyde V., and Whelpton, P.K. 1953: "Resume of the Indianapolis Study of Social and Psychological Factors Affecting Fertility in Population Studies", Vol. 7.

TABLE 9.3

Modernity value orientations and desired ideal and actual family size

Level of modernity	Ugala			Ghagga		
	Av. ideal family size	Mean no. of live-births	No. & % of respondents	Av. ideal family size	Mean no. of live-births	No. & % of respondents
Traditional	3.14	5.19	120 (51.06)	3.33	5.39	91 (33.33)
Moderately modern	2.90	4.32	86 (36.60)	3.23	4.51	119 (43.59)
Modern	3.17	4.83	29 (12.34)	2.70	4.20	63 (23.08)
Total	3.06	4.82	235 (100.00)	3.73	4.73	273 (100.00)

	Ghudani Kalan			Total		
	Av. ideal family size	Mean no. of live-births	No. & % of respondents	Av. ideal family size	Mean no. of live-births	No. & % of respondents
Traditional	3.17	4.97	107 (35.69)	3.20	5.17	308 (39.36)
Moderately modern	3.10	3.85	124 (41.33)	3.09	4.21	321 (40.72)
Modern	2.30	3.30	69 (23.00)	2.52	3.93	161 (19.32)
Total	2.90	4.12	300 (100.00)	3.02	4.53	808 (100.00)

3. Rainwater, Lee 1960: And the Poor Get Children: Sex, Contraception and Family Planning in the Working Class, Chicago: Quardrangle Books.
4. Mukherjee, B.N. 1979: Prediction of Family Planning and Family Size from Modernity Value Orientations of Indian Women. Papers of the East-West Population Institute, Hawaii, No. 61.
5. Pareek, Udai and V. Kothandapani 1969: "Modernization and Attitude towards Family Size and Family Planning: Analysis of Some Data from India", Social Biology, Vol. 16, 1.
6. George, Molly Kutty: "Application of Mahalanbois D^2 in a Study of Tradition and Fertility", Journal of Family Welfare, Vol. 19, 1.
7. Singh, K.P. 1982: "Modernity Value Orientations, Fertility and Family Planning", The Journal of Family Welfare, Vol. 20, 2.

10

FAMILIAL VARIABLES AND FERTILITY

In the present chapter, it is proposed to examine the role of familial variables in affecting fertility behaviour. It has been observed that besides individual variables like age, educational level and other socio-cultural characteristics, reproductive behaviour is also influenced by some other variables connected with the individuals' relationship with the institution of family. In the words of Pareek and Rao (1974) there are some extrinsic variables which affect the acceptance of family planning by restricting the freedom of action of the individual, pressuring him/her to behave in certain ways, and by motivating him/her either to accept or resist family planning. These are the variables related to the family, community or society.

Under familial variables are, thus, included the type of family structure, the interspouse communication, desired ideal family size, reasons for having sons and daughters and advantages and disadvantages of a large family size.

Family Structure and Fertility

The fertility level of a community or society is greatly influenced by its family structure as family being the most universal and permanent institution of every society has enormous influence on the life of the individual.

There are two points of view regarding the relationship between family structure and fertility behaviour. According to one, the joint family promotes higher fertility than the nuclear families. It is because the joint or extended family provides a congenial environment for higher fertility as the economic cost of and burden of rearing and bearing of children are shared by other members of the family. In such families marriage is generally universal and occurs early because it is not necessary for a married couple to support themselves and their children immediately after marriage.

Compared to this, in nuclear families couples have to bear the cost of rearing and bearing of children themselves. Also, the couples have fewer obligations to other family members as well as fewer expectations from them and are, therefore, under less pressure to fulfil the expectations of higher fertility norm prevailing in the joint family.

According to the other point of view, the joint family operates to depress fertility. For example, privacy is greater in a nuclear than in a joint family. The freedom of a couple to meet in privacy is, to a large extent, curtailed in a joint family by the traditional pattern of behaviour enforced by the elders of the family. Apart from that, the traditional taboos of sexual intercourse on religious and festive days are more rigorously enforced because of the presence of elders in the family.

The proponents of these views have, thus, identified some of the intermediate variables like age at marriage, coital frequency and interpersonal communication through which family structure influences fertility behaviour.

Studies dealing with family type and fertility are not many in India but whatever these studies are, most of them show that the average number of children born, when standardised for mother's age, duration of marriage and socio-economic conditions, are fewer for extended families than the nuclear families.

Bebarta (1976) from his study of 567 women belonging to joint families and 327 nuclear families from six Delhi villages, concluded that the nuclear type had higher fertility than the joint type in each age group and for the overall sample. The average completed family size for the joint family was 8.06 and 8.28 for the nuclear family. Karkal (1975) concluded on the basis of data collected from 27 villages in Maharashtra, on the other hand, that extended families had larger average number of live-births than nuclear families when standardised for duration of age at effective marriage and age at marriage of the women, but she found considerable difference in the age at effective marriage, particularly, in the "up to 5 years" group. The differences here are small and Karkal considers that the results of her study may differ due to regional factors. Lorimer, Davis and other (1954) also maintained that high fertility in tradional cultures was caused by extended family systems. Their argument is based on extended family's ability to extend financial support and help in child rearing to the newly married couple.

Since there is no clear evidence of a fertility differential according to family type and this relationship may vary from one community to another, it is worthwhile examining this relationship in a rural community where majority of the population is engaged in agriculture and live in joint families. But before we look into this relationship, it becomes necessary to define these two terms as used in the present study. By nuclear family is meant here a family consisting of husband, wife and their unmarried children. Joint family has been broadly defined to include any other member or relative, whether married or unmarried.

It may be mentioned here that, on the whole, 52 per cent of the respondents were living in joint households and the remaining 47 per cent of them were living in nuclear families. But for each village separately, Ghudani Kalan had the highest percentage of joint families, i.e., 62 per cent, followed by Ugala, i.e., 49 per cent and Ghagga had 47 per cent of the respondents living in joint families. Table 10.1 gives the distribution of respondents by their family type and mean number of live-births.

The data in Table 10.1 clearly reveal that for the total sample as well as for each village separately, mean number of live-births were higher in nuclear as against the joint families. The difference between the two was well marked for the total sample as also for Ghagga, although in the other two villages the difference was rather small.

The data further show that among the respondents of nuclear families, mean number of live-births were observed to be higher in Ghagga and comparatively lower in Ghudani Kalan, but, in joint families, the highest number of live-births were found in Ugala and the lowest in Ghudani Kalan.

Although, on the whole, mean number of live-births were observed to be higher in nuclear than in joint families, still in

TABLE 10.1

Mean number of live-births by family type

Family type	Ugala		Ghagga		Ghudani Kalan		Total	
	Mean number of live-births	Number & % of respondents	Mean number of live-births	Number & % of respondents	Mean number of live-births	Number & % of respondents	Mean number of live-births	Number & % of respondents
Single member family	—	—	—	—	3.00	6 (2.00)	3.00	6 (1.24)
Nuclear family	4.94	119 (50.64)	5.16	144 (52.75)	4.40	107 (35.67)	4.87	370 (47.15)
Joint family	4.72	116 (49.36)	4.23	129 (47.25)	4.01	187 (62.33)	4.27	432 (51.61)
Total	4.82	235 (100.)	4.73	273 (100.)	4.12	300 (100.)	4.53	808 (100.)

order to eliminate the influence of any chance factor, an attempt was made to examine this relationship with the duration of marriage. Table 10.2 gives mean number of live-births by duration of marriage and family type.

The table shows that in the shorter duration group, i.e., 19 or less than 19 years, nuclear families in the total sample as well as in each village separately had higher number of live-births than the joint families but reverse was true in the longer duration of marriage. In the longer duration of marriage, i.e., 20 years and more, where women were either near the completion of their reproductive span or had already completed it, with the exception of Ghagga, in the total sample as well as in Ugala and Ghudani Kalan, mean number of live-births were higher in joint than in nuclear families. It is, thus, clear that the overall fertility was comparatively higher for nuclear family but when controlled for marriage duration, it was higher for joint families.

Since it has been assumed that this relationship operates through certain intermediate variables, an attempt has been made to examine the role of age at marriage, incidence of infant and child deaths, use of family planning methods and the extent of interspouse communication in fertility differentials by family type. Table 10.3 gives mean number of live-births according to age at marriage, percentage of child deaths, users and non-users of family planning methods and interspouse communication in joint and nuclear families.

Contrary to the popular opinion that marriage generally occurs early in joint families as it is not necessary for a married couple to support themselves and their children immediately after marriage, the data in the above table show that mean age at marriage was higher in joint as compared with the nuclear families in the total sample as also in each village separately. Therefore, if respondents in nuclear families, on the whole, had higher number of live-births, their age at marriage was lower than the respondents in joint families.

As far as the incidence of infant and child deaths was concerned tbe data show that in the total sample the percentage of child deaths was little higher among respondents in joint families though the difference was small. However, no uniform

TABLE 10.2 (a)

Mean number of live-births, marriage duration and family type

Family type	Ugala					Ghagga				
	Marriage duration					Marriage duration				
	0—19 Yrs		20+ Yrs			0—19 Yrs		20+ Yrs		
	Mean no of live-births	No. & % of respon-dents	Mean no. of live-births	No. & % of respon-dents	Total	Mean no. of live-births	No. & % of respon-dents	Mean no. of live-births	No. & % of respon-dents	Total
Single member family	—	—	—	—	—	—	—	—	—	—
Nuclear family	4.39	46 (39.66)	5.34	73 (61.34)	119 (50.64)	3.23	80 (54.79)	7.63	64 (50.39)	144 (52.75)
Joint family	2.23	70 (60.34)	8.40	46 (38.66)	116 (49.36)	2.08	66 (45.21)	6.46	63 (49.61)	129 (47.25)
Total	3.09	116 (100.00)	6.52	119 (100.00)	235 (100.00)	2.71	145 (100.00)	7.05	127 (100.00)	273 (100.00)

TABLE 10.2 (b)

Mean number of live-births, marriage duration and family type

Famliy type	Ghudani Kalan — Marriage duration — 0–19 Yrs		20+ Yrs		Total	Total sample — Marriage duration — 0–19 Yrs		20+ Yrs		Total
	Mean no. of live-births	No. & % of respondents	Mean no. of live-births	No. & % of respondents		Mean no. of live-births	No. & % of respondents	Mean no. of live-births	No. & % of respondents	
Single member family	—	—	3.00	6 (3.75)	6 (2.00)	—	—	3.00	6 (1.48)	6 (0.74)
Nuclear family	3.17	46 (32.86)	5.28	61 (38.13)	107 (35.67)	3.52	172 (42.79)	6.06	198 (48.77)	370 (45.79)
Joint family	2.54	94 (67.14)	5.49	93 (58.12)	187 (62.33)	2.31	230 (57.21)	6.46	202 (49.75)	432 (53.47)
Total	2.75	140 (100.00)	5.32	160 (100.00)	300 (100.00)	2.83	402 (100.00)	6.21	406 (100.00)	808 (100.00)

TABLE 10.3 (a)

Mean number of live-births by women's age at marriage, percentage of child deaths, interspouse communication and users of contraception

	Ugala				Ghagga			
	Single member family	Nuclear family	Joint family	Total	Single member family	Nuclear family	Joint family	Total
Mean no. of live-births	— (—)	4.94 (119)	4.72 (116)	4.82 (235)	— (—)	5.16 (144)	4.23 (129)	4.73 (273)
Mean age at marriage	—	17.36	17.84	17.70	—	17.40	18.03	17.86
Percentage of children dead	(—)	18.80 (64)	14.31 (81)	12.78 (145)	(—) (—)	14.92 (111)	12.79 (70)	14.02 (181)
Interspouse communication								
Communicators	— (—)	48 (40.34)	49 (42.24)	97	— (—)	68 (47.22)	53 (44 96)	126
Non-Communicators	— (—)	71 (59.66)	67 (57.76)	138	— (—)	76 (52.78)	71 (55.04)	147
Users of Contraception								
Users	— (—)	47 (39.50)	38 (32.76)	85	— (—)	59 (40.97)	41 (31 78)	100
Non-users	— (—)	72 (60.50)	78 (67.24)	150	— (—)	85 (59.03)	83 (68 22)	173

TABLE 10.3 (b)

Mean number of live-births by women's age at marriage, percentage of child deaths, interspouse communication and users of contraception

	Ghudani Kalan				Total sample			
	Single member family	Nuclear family	Joint family	Total	Single member family	Nuclear family	Joint family	Total
Mean no. of live-births	3.00	4.40	4.01	4.12	3.00	4.87	4.27	4.53
Mean age at marriage	(6) 19.00	(107) 17.45	(187) 17.96	(300) 17.80	(6) 19.00	(370) 17.52	(432) 17.95	(808) 17.76
Percentage of children dead	11.11 (2)	19.79 (93)	17.51 (131)	18.28 (226)	11.11 (2)	14.87 (268)	15.31 (282)	15.07 (552)
Interspouse communication								
Communicators	—	29 (27.10)	53 (28.34)	82	—	145 (39.19)	160 (37.04)	305
Non-Communicators	6 (100.00)	78 (72.90)	34 (71.66)	218	6 (100.00)	225 (60.81)	272 (62.96)	503
Users of contraception								
Users	—	33 (30.84)	44 (23.5)	77	—	139 (37.57)	123 (28.47)	262
Non-users	16 (100.00)	74 (69.16)	143 (76.47)	223	6 (100.00)	231 (62.43)	309 (71.53)	546

trend was observed in all the villages. The percentage of child deaths was higher among respondents in nuclear families in Ghagga and Ghudani Kalan but opposite was the case in Ugala. Thus, in the absence of any uniform trend much reliability can be placed on this variable.

There are few studies which explore the problem of association between family type and use of contraception. Some of these studies have reported that nuclear families are more easily motivated for contraception than the joint families. Bebarta (1967) has reported that more women belonging to nuclear families than joint families are aware of family planning. Use of contraception is also on a comparatively smaller scale in joint families because they lack privacy. Although the data in the present study also reveal more women using contraception in nuclear as compared to joint families, yet the influence of contraception on fertility is a function of various other variables, therefore, there does not seem to be any association between the two in the present sample.

Another factor that impinges on the family type and contraception dimension is the interpersonal or interspouse communication. The joint family has often been cited as a barrier in effective husband wife communication resulting in decreased use of contraception. Poffenberger's study (1968) of a village in Gujrat revealed that in the traditional joint family setting, effective communication seldom took place between the husband and wife, preventing effective practice of family planning.

The data in the above table show that for the total sample as well as for Ghagga, more women in nuclear families had communication with their husbands regarding the number of children they would like to have. But reverse was the case in Ugala and Ghudani Kalan, though on the whole, the difference between the nuclear and joint families was very small. Thus, here, too, in the absence of any clear trend, no generalization could be made about the influence of family type on interspouse communication.

Lastly, the number of children considered ideal for a family is also expected to be influenced by the family type. It is maintained that large family size is valued more in joint than in nuclear families because of the family labour and old age security. Table 10.4 give desired ideal family size by family type.

Contrary to our expectations the data reveal that number of children considered ideal for a family was higher among respondents in nuclear as compared with joint families for the total sample as well as for each village, although the difference was rather small.

Thus, on the whole, nuclear families had higher number of live-births although within the longer duration of marriage joint families had higher number of live-births. But the variations in the number of live-births between the two types of families, on the whole, indicate weak differentials.

Interspouse Communication and Fertility

Interspouse communication or communication between husband and wife on matters relating to family planning is an important factor influencing the decision to adopt contraception in order to control fertility behaviour, as the level of communication may indicate trust and the level of joint decision making in the particular family. A report of the Economic and Social Commission for Asia and Pacific (ESCAP 1974) indicates that the existence of open channels of communication between husband and wife is an important factor in the adoption of family limitation practices. Stressing the importance of international variables, Tien (1968) points out that in matters of fertility decision, patterns and processes of interaction between husband and wife can undoubtedly have major consequence for family size.

Thus, the fact that communication between husbands and wives on matters relating to family planning is an important factor in the successful practice of contraception control is evident from several studies (Hill, Stycos 1959, Misra 1966, Jolly 1976 and Poffenberger 1968). All these studies suggest that the communication between husband and wife encourages the use of contraception and thus depresses fertility. In the following pages an attempt has been made to examine the effect of interspouse communication on the mean number of live-births and the acceptance of contraception.

TABLE 10.4

Ideal family size Vs. family type

Family type	Ugala		Ghagga		Ghudani Kalan		Total	
	Mean number of live-births	Number & % of respondents	Mean number of live-births	Number & % of respondents	Mean number of live-births	Number & % of respondents	Mean number of live-births	Number & % of respondents
Single member family	—	—	—	—	3.00	5 (1.82)	3.00	5 (0.73)
Nuclear family	3.07	88 (45.13)	3.23	108 (50.00)	2.99	97 (35.27)	3.07	293 (42.71)
Joint family	3.03	107 (54.87)	3.03	108 (50.00)	2.91	173 (62.91)	2.98	388 (56.56)
No response	—	40	—	57	—	25	—	112
Total	3.00	235 (100.)	3.13	273 (100.)	2.70	300 (100.)	3.02	808 (100.)

It may be mentioned here that in order to understand the level of interspouse communication two questions were asked from the respondents. The first question was whether they have ever talked to their husbands about the number of children they would like to have, and, the second referred to their discussion regarding the use of family planning methods. Since the couples may vary in terms of their level of communication they have between them, the communicators or those who had discussed with each other were divided into frequent and occasional communicators. But for the purpose of analysis these two categories of respondents have been put into one category as the number of respondents in each cell when analysed separately becomes too small to make any valid generalization. Thus, the respondents have been divided into communicators and non-communicators.

In the total sample only 37.8 per cent of the respondents were communicators but this percentage varied for each village as the highest percentage of communicators, i.e., 46.5 per cent, were found in Ghagga and the lowest percentage, i.e., 27.6 per cent, in Ghudani Kalan. Table 10.5 gives the distribution of communicators and non-communicators and mean number of live-births.

It is clear from Table 10.5 that those respondents who had discussed with their husbands regarding the number of children they would like to have had lesser number of live-births as compared to those respondents who were non-communicators or who had never discussed it, in the total sample as well as in each village, and the difference between the two categories of respondents was quite significant.

Since communication between the husband and the wife encourages the use of contraception, it is expected that those who had communication between themselves regarding the number of children they would like to have should also adopt contraception to control the size of their families or to achieve the desired number of children. Therefore, Table 10.6 gives the distribution of communicators and non-communicators according to the use of contraception.

Table 10.6 shows that among communicators almost fifty per cent of them for the total sample as well as for every village,

TABLE 10.5

Mean number of live-births among communicators and non-communicators

Category of respondents	Ugala		Ghagga		Ghudani Kalan		Total	
	Mean no. of live-births	No. & % of respondents	Mean no. of live-births	No. & % of respondents	Mean no. of live-births	No. & % of respondents	Mean no. of live-births	No. & % of respondents
Communicators	4.17	96 (40.85)	4.05	127 (46.52)	3.06	83 (27.67)	3.82	306 (37.87)
Non-communicators	5.28	139 (59.15)	5.32	146 (53.48)	4.52	217 (72.33)	4.96	502 (62.13)
Total	4.82	235 (100.0)	4.73	273 (100.0)	4.12	300 (100.0)	4.53	808 (100.0)

TABLE 10.6

Use of contraception among communicators and non-communicators regarding the number of children they would like to have

Category of respondents	Ugala			Ghagga		
	Users No. & %	Non-users No. & %	Total	Users No. & %	Non-users No. & %	Total
Communicators	47 (48.96)	49 (51.04)	96 (100.)	64 (50.39)	63 (49.61)	127 (100.)
Non-communicators	38 (27.34)	101 (72.66)	139 (100.)	36 (24.66)	110 (75.34)	146 (100.)
Total	85 (36.17)	150 (63.83)	235 (100.)	100 (36.63)	173 (63.37)	273 (100.)

Category of respondents	Ghudani Kalan			Grand Total		
	Users No. & %	Non-users No. & %	Total	Users No. & %	Non-users No. & %	Total
Communicators	39 (46.99)	44 (53.01)	83 (100.)	150 (49.02)	156 (50.98)	306 (100.)
Non-communicators	38 (17.51)	179 (82.49)	217 (100.)	112 (22.31)	390 (77.60)	502 (100.)
Total	77 (25.67)	223 (74.33)	300 (100.)	262 (32.43)	546 (67.57)	808 (100.)

were either using family planning methods at the time of this survey, or, had used these methods at some stage, compared to nearly one fifth of the non-communicators who were using some contraception.

This indicates that the communication between the husband and wife regarding the number of children that they would like to have had resulted in greater adoption of contraception and lesser number of children as compared to non-communicators.

Another aspect of communication which is more intimate in nature was whether the respondents had ever talked to their husbands regarding the use of family planning methods. Compared to the general discussion regarding the number of children they would like to have, the percentage of respondents who had discussed the use of family planning methods with their husbands was rather small, i.e., only 22 per cent in the total sample with very little variations in different villages. Since the discussion of family planning methods also implies adoption of these methods it is expected that thoes who had discussed this matter would use these methods more frequently than those who did not discuss it. Table 10.7 gives the distribution of respondents according to the communication regarding the use of family planning methods and the mean number of live-births.

The table indicates the influence of communication regarding the use of contraception on the number of live-births in every village although the difference between the two categories of respondents was not very sharp.

It is further expected that the adoption of contraception would be higher among these respondents who had discussed the use of contraception among themselves as against those who had never discussed it. Table 10.8 gives the extent of the use of contraception among communicators and non-communicators.

Table 10.8 clearly reveals that the practice of contraception was comparatively higher, i.e., nearly ninety per cent, for the total sample as well as for every village among communicators, whereas, it was almost negligible among non-communicators. This, therefore, supports the contention that the greater the communication between the husband and the wife, the

TABLE 10.7

Mean number of live-births among communicators and non-communicators regarding the use of contraception

Category of respondents	Ugala		Ghagga		Ghudani Kalan		Total	
	Mean no. of live-births	No. & % of respondents	Mean no. of live-births	No. & % of respondents	Mean no. of live-births	No. & % of respondents	Mean no. of live-births	No. & % of respondents
Communicators	4.70	63 (26.81)	4.52	53 (19.41)	4.03	64 (21.33)	4.41	180 (22.28)
Non-communicators	4.87	172 (73.19)	4.78	220 (80.59)	4.15	236 (78.67)	4.57	628 (77.72)
Total	4.82	235 (100.)	4.73	273 (100.)	4.12	300 (100.)	4.53	808 (100.)

TABLE 10.8

Use of contraception among communicators and non-communicators regarding the use of contraception

Category of respondents	Ugala			Ghagga		
	Users No. & %	Non-users No. & %	Total	Users No. & %	Non-users No. & %	Total
Communicators	59 (93.65)	4 (6.35)	63 (100.)	47 (88.68)	6 (11.32)	53 (100.)
Non-communicators	26 (15.12)	146 (85.47)	172 (100.)	53 (24.09)	167 (72.61)	220 (100.)
Total	85 (36.17)	150 (63.83)	235 (100.)	100 (36.63)	173 (63.37)	273 (100.)

Category of respondents	Ghudani Kalan			Grand Total		
	Users No. & %	Non-users No. & %	Total	Users No. & %	Non-users No. & %	Total
Communicators	58 (90.63)	6 (8.37)	64 (100.)	164 (91.11)	16 (8.89)	180 (100.)
Non-communicators	19 (8.05)	217 (91.95)	236 (100.)	98 (15.61)	530 (84.39)	628 (100.)
Total	77 (25.67)	223 (74.33)	300 (100.)	262 (32.43)	546 (67.57)	808 (100.)

greater the adoption of family planning methods and the greater the adoption, the smaller the family size.

It is further expected that the idea of a desired ideal family size would vary among communicators and non-communicators. Table 10.9 gives the distribution of communicators and non-communicators about family planning methods and the desired ideal family size.

It is again clear from Table 10.9 that the Communicators, i.e., those who had discussed the use of family planning methods, with their husbands had specified lesser number of children for an ideal family as compared to non-communicators for the total sample as well for every village.

It may, therefore, be concluded from the above that, since the decision to have a certain number of children is mainly that of the couple, the free communication between them results in greater adoption of contraception and smaller family size.

Ideal Family Size

The norms of ideal family size exist in every society. These norms are generally influenced by the need for children, values attached to children and preference for a particular sex. This, in turn, influences an individual's attitudes towards related aspects of family formation. Therefore, in the following paragraphs an attempt has been made to analyse the desired ideal family size and its relationship with the actual family size. An attempt would also be made to determine whether socio-economic characteristics of the respondents have any influence on their perception of the desired ideal family size.

The desired ideal family size is indicative of the women's acceptance of the norm of ideal family size. They may achieve this ideal or may not, but it does influence the intensity of motivation for adopting family planning devices. Table 10.10 gives the desired ideal family size along with the sex of the child.

The table indicates that large majority of the respondents for the total sample as well as for every village preferred three children for an ideal family and this was followed by those who preferred two and four children for an ideal family. The percentage of respondents who specified more than four children

TABLE 10.9

Ideal family size among communicators and non-communicators

Category of respondents	Ugala		Ghagga		Ghudani Kalan		Total	
	Mean number of live-births	No. & % of respondents	Mean number of live-births	No. & % of respondents	Mean number of live-births	No. & % of respondents	Mean number of live-births	No. & % of respondents
Communicators	2.98	58 (29.74)	2.88	48 (22.22)	2.83	63 (22.91)	2.89	169 (24.64)
Non-communicators	3.12	137 (70.26)	3.20	168 (77.77)	2.92	212 (77.09)	3.06	517 (75.36)
Total	3.06	195 (100.)	3.13	216 (100.)	2.90	275 (100.)	3.02	686 (100.)

TABLE 10.10

Distribution of respondents by their ideal family size

Ideal number of children	Ugala		Ghagga		Ghudani Kalan		Total	
	Number of respondents	Percentage	Number of respondents	Percentage	Number of respondents	Percentage	Number of respondents	Percentage
1 child (1 boy)	2	1.03	2	0.93	—	—	4	0.58
2 children (1 boy 1 girl/2 boys 2 girls)	24	12.31	54	25.00	84	30.55	162	23.62
3 children (2 boys 1 girl)	136	69.74	92	42.59	150	54.55	378	55.10
4 children (2 boys 2 girls/3 boys 1 girl)	25	12.82	63	29.17	31	11.27	119	17.35
5 children and more (3 boys 1 girl/4 boys/4 boys, 1 girl 4 boys 4 boys 2 girls	8	4.10	5	2.31	10	3.63	23	3.35
Total	195	100.00	216	100.00	275	100.00	686	100.00
Average ideal family size	3.06		3.03		2.90		3.02	

for an ideal family size was rather very small in each village. In terms of average ideal family size it was 3.02 for the total sample population, whereas, it was 3.06, 3.13 and 2.90 children for Ugala, Ghagga and Ghudani Kalan, respectively. Thus, on the whole, in all the villages the general preference was for three to four children and inter-village variations were not very significant.

It is generally observed that when people mention ideal family size, they have specific sex ratio in mind. In our society because of dowry problem and also because sons are treated as an old-age security, there is always greater preference for sons and the data in Table 10.10 are in conformity with it. Although large majority of the respondents mentioned children of both the sexes but the preference was for more sons than daughters. Majority of the respondents preferred only one daughter but the number of respondents who did not mention any daughter was rather small.

The preference for an ideal family is, to a great extent, influenced by the number of children already living in the family. Various Indian studies show that there is a wide gap between the preferred family size and the actual family size. Many of these studies show that those respondents who had smaller number of children tend to consider their family as an ideal one, but those who have large families, also consider that an ideal family should consist of small number of children. Table 10.11 gives the desired ideal family size in accordance with the actual number of children in the family.

Table 10.11 clearly reveals variations in the desired ideal family size by the actual number of children for all the villages. Those respondents who had one to two children specified little more than two children for an ideal family, whereas among those who had three to four children, large majority of them considered their family size as the ideal one. But among those who had more than four children stated four or less than four children for an ideal family.

The data further reveal that inter-village variation in the ideal family size were not very significant which is indicative of the fact that in all these rural areas, the respondents feel that the ideal family should consist of three to four children.

TABLE 10.11

Ideal family size by the actual family size

Actual number of children	Ugala		Ghagga		Ghudani Kalan		Total	
	Average ideal family size	Number & % of respon-dents	Average ideal family size	Number & % of respon-dents	Average ideal family size	Number & % of respon-dents	Average ideal family size	Number & % of respon-dents
0	3.17	6 (3.00)	3.11	18 (8.33)	2.56	9 (3.27)	2.97	33 (4.81)
1—2	2.58	42 (21.54)	2.78	54 (25.00)	2.59	71 (11.28)	2.67	167 (24.34)
3—4	3.08	51 (26.15)	3.13	44 (20.37)	2.72	85 (30.91)	2.92	180 (26.24)
5–6	3.16	56 (28.72)	3.26	42 (19.49)	3.13	63 (22.91)	3.17	161 (23.47)
7 and more	3.27	40 (20.51)	3.38	58 (26.85)	3.40	47 (17.09)	3.36	145 (21.23)
Average ideal family size	3.06	195 (100.)	3.13	216 (100.)	2.90	275 (100.)	3.02	686 (100.)

Age of the women is another important factor which might influence the perception of an ideal family size. It is expected that younger women, who perhaps are more educated and are exposed to the family planning programme, would prefer smaller number of children for an ideal family size as compared to older women. Table 10.12 gives the desired ideal family size by the present age of the respondents.

Table 10.12 shows that, on the whole, the younger women specified lesser number of children for an ideal family size as compared to older women, although within each village, there were some inconsistencies. But in spite of these inconsistencies within the same age groups, the older women, above the age of 45 years, stated higher number of children for an ideal family as compared with other age groups.

Socio-Economic Variables and Ideal Family Size

Since sociological variables like caste, religion and modernity value orientations have been analysed in other chapters, here only some of the socio-economic variables would be taken into account.

So far as the educational level of the respondents was concerned, on the whole, illiterate or just literate respondents mentioned higher number of children for an ideal family as compared to educated respondents. Similar trend was observed when the husband's educational level was taken into consideration.

It was further observed that working women specified lesser number of children for an ideal family as compared to non-working ones. However, the occupation of the husband did not indicate any clear trend regarding the ideal family size.

The economic status of the respondents when analysed in terms of level of living clearly revealed an inverse relationship between the two. An important point noted was the respondents whose husbands were land owners mentioned smaller number of children for an ideal family as against those who did not own any land.

Thus, on the whole, the desired ideal family size varied with the socio-economic conditions of the respondents.

TABLE 10.12

Ideal family size by women's present age

Present age in years	Ugala		Ghagga		Ghudani Kalan		Total	
	Average ideal family size	Number & % of respondents	Average ideal family size	Number & % of respondents	Average ideal family size	Number & % of respondents	Average ideal family size	Number & % of respondents
15—24	3.00	34 (17.44)	2.96	47 (21.76)	2.69	29 (10.55)	2.90	110 (16.03)
24—34	3.00	65 (33.33)	2.95	76 (35.19)	2.77	82 (29.82)	2.90	223 (32.51)
35—44	2.90	41 (21.03)	3.39	38 (17.59)	2.75	67 (24.36)	2.97	146 (21.28)
45 and above	3.27	55 (28.20)	5.14	55 (25.46)	3.13	97 (35.27)	4.21	207 (30.17)
Average ideal family size	3.06	195 (100.)	3.13	216 (100.)	2.90	275 (100.)	3.02	686 (100.)

Desire for More Children

Another important variable that impinges on the fertility and contraceptive behaviour in the family is the desire to have more children than the number they already have. Therefore, the respondents were asked whether they would like to have any more children, and if so, their preference for the sex of the child.

The data indicate that only one fourth of the respondents in the total sample expressed a desire to have additional children. However, this percentage was a little higher (29.0 and 30.0) for Ugala and Ghagga and lower (21.3 per cent) for Ghudani Kalan. Further, in terms of the sex of the desired child, it was noted that little more than fifty per cent of the respondents in the total sample desired to have only a male child, whereas, the remaining ones, with few exceptions, desired to have a male along with a female child.

It is generally observed that the desire for more children is closely linked to the number of children the women already have, the sex of the child or children as well as the age of the women. If a woman has just one or two living children and has not achieved the desired ideal family size, she is likely to desire more children. Similarly, if she has all or more children of one sex, she would desire to have more children to obtain the desired sex ratio in the family. Along with these two considerations is the age of the woman. In fact, in spite of these two considerations a woman would like to have another child only when she is young enough to bear and rear the child. Table 10.13 gives the distribution of respondents according to their desire to have an additional child or children by the number of children they already have.

It is clear from Table 10.13 that nearly 60 per cent of the respondents in every village with one to two living children desire to have an additional child. The percentage of respondents with more than three living children, desiring additional children, was rather small, indicating thereby that it is either those women who have not yet achieved the desired ideal family size or desired sex ratio who were keen to have additional children.

TABLE 10.13

Desire for additional children Vs. number of living children

Number of living children	Ugala		Ghagga		Ghudani Kalan		Total	
	No. of respon-dents	Percent-age	No. of respon-dents	Percent-age	No. of respon-dents	Percent-age	No. of respon-dents	Percent-age
0	6	8.82	18	22.22	8	12.50	32	15.02
1-2	39	57.35	43	53.09	41	64.06	123	57.75
3-4	19	27.94	14	17.28	13	20.31	46	21.60
5-7+	4	5.88	6	7.41	2	3.13	12	5.63
Total	68	99.99	81	100.00	64	100.00	213	100.00

Since the age of women is an important factor that influences the desire for more children. Table 10.14 gives the distribution of respondents according to their desire for more children and present age of the respondents.

Table 10.14 clearly reveals that nearly 80 per cent of the respondents who wanted to have more children were below the age of 29 in every village and the percentage of respondents desiring more children was very small beyond the age of 35 years.

Thus on the whole it was only young women with one to two living children, who expressed desire for additional children.

Reasons for Having Sons and Daughters

Children are valuable to parents since they provide social, economic and psychological satisfaction. But parents generally do not attach similar values to sons and daughters. Their preferences related to the gender of children are largely a reflection of sex role differences in the community of society which, in turn, have an effect on the differential perception of children by men and women.

Parents generally have adverse needs and these are fulfilled by having children of each sex. Sons are generally valued for long term economic benefits, including old age security. They are also wanted for their role in continuing the family lineage. Daugthers, on the other hand, tend to be valued for their help they tender in household work and for religious reasons. In the present study also, with a view to understand the values that these respondents attach to sons and daughters, they were asked, "Some people think that the reasons for the desire to have sons and daughters are the same, whereas, there are other people who think that the reasons for having sons and daugthers are different. What are your views?" Table10.15 gives respondent's views regarding sons and daughters.

The data in Table 10.15 reveal that large majority of the respondents in the total sample as also in every village, put forward different reasons for having sons and daugthers. However, inter-village variations in terms of these reasons were quite significant as this percentage was 86 for Ugala, 72 for

TABLE 10.14

Desire for additional children Vs. present age of the women

Present age of women	Ugala		Ghagga		Ghudani Kalan		Total	
	No. of respon-dents	Percent-age	No. of respon-dents	Percent-age	No. of respon-dents	Percent-age	No. of respon-dents	Percent-age
15-19 yrs.	4	5.88	14	7.28	3	4.69	21	9.86
20-24 yrs.	30	44.12	28	34.57	22	34.38	80	37.56
25-29 yrs.	18	26.47	23	28.40	24	37.50	65	30.53
30-34 yrs.	12	17.65	13	16.05	8	12.50	33	15.49
35 and above	4	4.88	3	3.70	7	10.93	14	6.57
Total	68	100.00	81	100.00	64	100.00	213	100.00

TABLE 10.15

Distribution of respondents by their reasons for having sons and daughters

Reasons	Ugala		Ghagga		Ghudani Kalan		Total	
	No. of respondents	Percentage	No. of respondents	Percentage	No. of respondents	Percentage	No. of respondents	Percentage
Same	26	11.06	90	32.97	78	26.00	194	24.01
Different	202	85.96	179	65.57	218	72.67	599	74.13
Can't say	7	2.98	4	1.46	4	1.33	15	1.86
Total	235	100.00	273	100.00	300	100.00	808	100.00

Ghudani Kalan and 65 for Ghagga. Considering the values which Indian parents attach to sons and daughters, it is at the same time heartening to note that nearly one fourth of the respondents in the total sample did not make any distinction between sons and daughters and stated that they attached equal value to children of both the sexes.

Contrary to these findings, one generally does not find very different reasons for having sons and daughters in the developed countries as children over there are mostly consumers rather than producers because there is hardly any scope for Child labour. Most the children remain in schools until their late teens. Moreover, parents do not depend upon them for old age security. Thus, parents over there are interested in children per se.

Although majority of the respondents in Table 10.15 did reflect traditional attitude towards sons and daughters, still it was expected that young and educated respondents would have a more liberal atttitude towards their desire for sons and daughters. But, contrary to our expectations, the data when analysed by the age and educational level of the respondents did not reveal any relationship indicating thereby that large majority of the rural women, whether young or old, highly educated or just literate, still held traditional ideas about sons and daughters.

Having mentioned same/different values for desiring sons and daughters, respondents were further asked to give reasons for their views.

Among those who attached the same values to sons and daughters, large majority of them in all the villages perceived no difference in sons and daughters and considered both the sexes equally necessary for a complete family. They felt that in the present age when girls are also going in for higher and technical education and have become career oriented, there is no distinction between the two. Some of them even felt that both the sexes are God's gift, both provide emotional satisfaction and both are equally helpful.

On the other hand, among those who attached different values to sons and daughters, large majority of them stated that sons are primarily desired for their role in the continuation of the family name. For them, continuation of family name is most essential, therefore, sons are valued for that. Some others also felt that sons are an old age security and they also contribute to the family finances. Few of them also mentioned moral support, more contacts and enhanced economic status along with these reasons.

Coming to the reasons for desiring daughters, majority of the respondents mentioned Kanyadan (gift of the daughter) as the most important reason. According to the Hindu tradition, the ritual of Kanyadan plays a significant role and it is often stated that Kanyadan is the most pious 'dan', hence one must perform it. Another important reason mentioned by large number of them was the help the daughters render in the household chores. Some of them also stated that daughters when young provide company to their mothers and when married increase the network of relationship which is beneficial to the family. Lastly, a few of them expressed concern for the society and stated that daughters necessary for the continuation of society, otherwise, who would provide girls for the boys if every one desires only sons.

Thus, on the whole, sons are mainly valued either for continuation of family name or as economic utility, whereas, daughters are valued for religious reasons as also for domestic help they provide to the family.

Advantages and Disadvantages of a Large Family

In order to understand whether a large family has any advantages or disadvantages and which advantages and disadvantages have salience for parents, the respondents were asked whether they perceived that a large family has any advantages or disadvantages and if so, the various types of advantages or disadvantages which they perceive in a large family. Table 10.16 gives the distribution of respondents according to their views about a large family.

It is clear from Table 10.16 that large majority of respondents indicated their concern for disadvantages as against advantages, as 77 per cent of them mentioned disadvantages as compared to only 7 per cent who emphasised only advantages. This, thus, clearly indicated that women were well aware of the limitations of a large family. Another 12 per cent of them felt

TABLE 10.16

Advantages and disadvantages of a large family

Name of village	Advantages of a large family		Disadvantages of a large family		Both advantages and disadvantages		Indifferent		Total
	No. of respondents	Percentage	No. of respondents	Percentage	No. of respondents	Percentage	No. of respondents	Percentage	
Ugala	23	9.79	178	75.74	23	9.79	11	4.68	235
Ghagga	33	12.09	208	76.19	22	8.06	10	3.66	273
Ghudani Kalan	4	1.33	236	78.67	52	17.33	8	2.67	300
Total	60	7.42	622	76.98	97	12.00	29	3.59	808

that a large family had both advantages as well as disadvantages depending upon the individual's position while the remaining 3 per cent could not make any specific response to the question.

Further, among those who referred only to the disadvantages, a large majority of them expressed concern due to financial considerations. They felt that it was difficult to support a large family as the cost of living was very high and it was difficult to fulfil the needs of several children. Almost one fourth of them stated that in a large family all the children could not be looked after properly, hence they got neglected. Some of them (8 per cent) also mentioned about family disputes and such other problems which often arose when there were too many people around. Few of them also referred to division of family property etc. It is interesting to observe that none of the respondents expressed any concern for restrictions on parent's freedom, health of the mother or other unpleasant aspects of having too many children. Compared to this, in a study on the value of children in Turkey (United Nations 1977) the most frequently mentioned disadvantage of children was "concern about the child", followed by "economic disadvantages" of children.

Regarding advantage of a large family, the most important factor mentioned by majortiy of them was economic utility as they felt that more children meant more earning hands and more earners meant more income for the family. Some of them also mentioned that in a large family, responsibility could be divided and work could be done without outside assistance. Here, too, none of them referred to such reasons as expansion of the self, affection and love and power etc. Contrary to this, in the Turkey study (United Nation 1977) the most salient advantage mentioned by a large majority of the respondents was found to be "primary group ties, affection and love followed by those who mentioned security in old age, stimulation, novelty and fun".

Thus, on the whole, these rural women in the present study mostly expressed their concern due to economic considerations in terms either of more earning hand or more financial burden and less about other aspects.

Lastly, it was observed that the respondents' perception of

the advantages or disadvantages of a large family were, to a great extent, influenced by their own actual family size as those who mentioned only the advantages of a large family had the highest (6 13 children) number of births, on the other hand, those who referred only to the disadvantages, had 4.26 number of births, whereas those who referred to both advantages as well as disadvantages depending upon the circumstances of each individual, had 5.0 live-births. But, on the whole, the majority of them showed their concern for disadvantages rather than advantages of a large family.

REFERENCES

1. Bebarta, P.C. 1967: "Attitude of Women towards Family Planning: A Study of Differences by Family Type in Six Villages of Delhi", Quarterly Journal of Indian Studies in Social Sciences, Bombay, 1(1).
2. Hill, R., Stycos, M.J, and Back, K.W. 1959: The Family and Population Control: A Puerto Rican Experiment in Social Change, Chapel Hill, University of North Carolina Press.
3. Karkal, M. 1975: "The Family and Fertility", Journal of Family Welfare, Vol. XXII, 1.
4. Lorimer, F., et al. 1957: Culture and Human Fertility, Paris, Unesco.
5. Pakrasi, K. and Malakar, C. 1957: "The Relationship between Family Type and Fertility", Milbank Memorial Fund Quarterly, 45(4).
6. Pareek, U. and Rao, T.V. 1974: A Status Study on Population Research in India, Vol. I, New Delhi, Tata McGraw Hill.
7. Poffenberger, Thomas 1969: Husband-Wife Communication in an Indian Village, New Delhi, Central Family Planning Institute.
8. United Nations 1974: Husband-Wife Communication and Practice of Family Planning Bangkok, Asian Population Studies Series, 16.
9. United Nations 1977: Demographic Transition and Socio-economic Development, Istambul, Proceedings of the United Nations Expert group meeting.

11

FAMILY PLANNING PRACTICE AND FERTILITY

In the present chapter an attempt has been made to ascertain the extent of the use of family planning practices and its influence on the fertility behaviour of the respondents. An attempt would also be made to examine some of the socio-economic correlates of contraceptive behaviour. The use of contraception greatly affects the fertility as it indicates the desire on the part of the women or the couple to use family planning methods either for spacing or for limiting the family size.

The data show that, on the whole, of the total number of respondents only 32 per cent of them were either making use of or had made use of some contraception. However, this percentage was not uniform for all the villages. It was 36.17 and 36.03 for Ugala and Ghagga, respectively, but it was only 25.67 per cent for Ghudani Kalan.

It was further observed that sterilization, whether male or female (31 per cent vasectomy and 34 per cent tubectomy), was the most commonly practised method of contraception by the respondents as its use was reported by almost two thirds of the respondents in the total sample and this proportion was more or less the same for all these villages. ORG Baroda (1972) study also reported that 65.6 per cent of the respondents had got themselves sterilized. This study found that the next most popular method was the use of IUD as it was reported by 32 per cent of the respondents. Nearly 4 per cent of them reported the use of other conventional methods also, whereas, none of them mentioned the use of pill or other clinical methods.

The use of sterilization by a large majority of the respondents indicates that these women adopted this method for limiting the family size rather than for spacing. This fact was further substantiated when the respondents were directly asked whether they were using these methods for spacing or for limiting the family size. Only 15 per cent of the respondents reported that they were using these methods for spacing, whereas, all others stated that contraception was being used for limiting the size of their families.

Since a large majority of the respondents resorted to contraception for family limitation, it was expected that the practice of family planning would be more wide spread among those women who had many children rather than those who had few of them. Table 11.1 gives the distribution of respondents according to pregnancy after which they started the use of these methods.

The table indicates that the highest percentage of respondents who were making use of contraception in each village were those women who had three to five pregnancies and this was followed by those respondents who adopted contraception after 5th pregnancy. The percentage of respondents using family planning methods after one or two pregnancies was quite small and none of them adopted any contraception soon after marriage or after the first pregnancy. This clearly shows that these women had adopted contraception only when they had produced certain number of children.

There were no significant variations in the use of contraception among these villages. However in Ghagga a comparatively higher proportion of women had adopted contraception after their fifth pregnancy.

It was further observed that, on the whole acceptors of contraception had higher number of live-births as against non-acceptors. Table 11.2 gives the mean number of live-births among users and non-users of contraception.

It is clear from Table 11.2 that the users of contraception had higher number of live-births as against non-users in the total sample as well as in each village separately. In a study of fertility differentials amongst contraceptors and non-contraceptors in India (Gandotra and Das 1980) it was also found that the standardised average number of children ever born per woman was 3.44 for the never users as against 4.05 for the

TABLE 11.1

Distribution of ever-users by the pregnancy after which the use of contraception started

Number of pregnancies	Ugala		Ghagga		Ghudani Kalan		Total	
	Number of respondents	Percentage	Number of respondents	Percentage	Number of respondents	Percentage	Number of respondents	Percentage
1—2	14	16.47	16	16.00	15	19.48	45	17.18
3—5	48	56.47	45	45.00	44	57.14	137	52.29
6 and above	23	27.06	39	39.00	18	23.38	80	30.53
Total	85	100.00	100	100.00	77	100.00	262	100.00

TABLE 11.2

Mean number of live-births among contraceptors and non-contraceptors

	Ugala		Ghagga		Ghudani Kalan		Total	
	Mean number of live-births	Number & % of respondents	Mean number of live-births	Number & % of respondents	Mean number of live-births	Number & % of respondents	Mean number of live-births	Number & % of respondents
Users	4.96	85 (36.17)	5.04	100 (36.63)	4.15	77 (25.67)	4.76	262 (32.43)
Non-users	4.77	150 (63.83)	4.55	173 (63.37)	4.11	223 (74.33)	4.42	546 (67.57)
Total	4.82	235 (100.00)	4.73	273 (100.00)	4.12	300 (100.00)	4.53	808 (100.00)

users. Similarly, in the study of GBFS (Rele and Patankar 1969) it was also noticed that the standardised average number of children ever born to the non-users and the ever-users of contraception was 2.48 and 3.21, respectively. The general explanation given for these differences was that the non-users included sterile and sub-fecund couples while the users were generally highly fecund. It was because of their higher fecundity that they accepted contraception.

It was further noted that fertility differentials were there not only in terms of the total number of live-births but also in terms of the sex of the child.

The data, on the whole, indicate that among users the number of male live-births was higher than the female live-births in these villages. However, this difference was more marked in case of sterilization as compared to other contraceptives indicating thereby that women adopted permanent contraception only when they had already produced 5-6 children, out of which more than half of them were male children. On the other hand, women who had comparatively smaller number of male children preferred to use temporary methods.

Fertility differentials were also observed among respondents according to the type of method used. Table 11.3 gives the mean number of live-births by the type of contraception used.

Mean Number of Live-births by the Type of Contraception Used

Table 11.3 shows that mean number of live-birth s was much higher for those women who either themselves or their husbands resorted to sterilization as against those respondents who made use of IUCD. Mean number of live-births was also much higher for those respondents who adopted conventional contraceptives but the number of such cases was too small to permit any generalization.

The table further shows that when given the use of same methods of family planning, mean number of live-births was comparatively smaller in village Ghudani Kalan, especially, for acceptors of sterilization and IUCD.

The age of the women appears to be another important factor related to the use of contraception as the users were mostly women of higher parity and it is expected that the use

TABLE 11.3

Mean number of live-births by the use of each method

Users	Ugala		Ghagga		Ghudani Kalan		Total	
	Mean no. of live-births	No. & % of respon-dents	Mean no. of live-births	No. & % of respon-dents	Mean no. of live-births	No. & % of respon-dents	Mean no. of live-births	No. & % of respon-dents
Sterilization	5.47	57 (67.06)	5.84	62 (62.00)	4.82	50 (64.94)	5.41	169 (64.50)
I.U.D.	3.87	23 (27.06)	3.50	36 (36.00)	2.63	24 (31.16)	3.35	83 (31.68)
Conventional methods	4.75	4 (4.71)	8.00	2 (2.00)	6.00	1 (1.30)	5.86	7 (2.67)
Other methods	2.00	1 (1.17)	—	—	5.00	2 (2.60)	4.00	3 (1.15)
Total	4.96	85 (100.00)	5.04	100 (100.00)	4.15	77 (100.00)	4.76	262 (100.00)

would be even higher for other ages. Table 11.4 gives the use of contraception according to the age of the respondents.

Table 11.4 clearly indicates that the highest percentage of acceptors of contraception were in the age group 30-39, followed by the age group 40 and above for the total sample as well as for each village separately. The percentage of acceptors was very low when they were young and below the age of 25 years. Inter-village variations were there but they were not very significant.

The age and parity of the women, in fact, go together as women generally complete their family size at higher ages, i.e., when they are above 30 and since they start producing children at a comparatively earlier age, by the time they reach their thirties, and adopt contraception, they had already given birth to fairly large number of children.

Since only one third of the respondents were the acceptors of contraception, they were asked whether they intended to use any methods in future. Of course this was not applicable to those respondents who had adopted permanent methods, i.e., sterilization, but the remaining respondents were divided into acceptors and non-acceptors.

From among the contraceptors of temporary methods, nearly one third of them mentioned that they would continue the use of contraception in future also, whereas the majority of them could not give any specific answer as more than half of them were above the age of 40, and they were not sure whether or not they would require the use of any methods in future. Only few of them replied in the negative as they were sure that they would not require these methods in future.

Among the non-acceptors, surprisingly, majority of them also replied in the negative. It is true that the use of contraception depends upon the needs of individual women. Some may not require the use of any method due to biological reasons, whereas, others may not require them as they have not given birth to the desired number of children or may not have desired sex ratio. Although little more than half of these women were above the age of 40, still even those who were below 40, did not like to use them in future without specifying any reasons for that. The number of respondents among these who intended

TABLE 11.4

Contraceptors and non-contraceptors by the present age of the women

Present age of the women in years	Ugala			Ghagga		
	Users No. & %	Non-users No. & %	Total	Users No. & %	Non-users No. & %	Total
15—19	1 (25.00)	3 (75.00)	4 (1.70)	1 (7.14)	13 (92.86)	14 (5.13)
20—29	15 (23.44)	49 (76.56)	64 (27.24)	20 (27.40)	53 (72.60)	73 (26.74)
30—39	42 (65.63)	22 (34.38)	64 (27.23)	39 (50.65)	38 (49.35)	77 (28.21)
40+	27 (26.21)	76 (73.79)	103 (43.83)	40 (36.70)	69 (63.30)	109 (39.92)
Total	85 (36.17)	150 (63.83)	235 (100.)	100 (36.63)	173 (63.37)	273 (100.)

Present age of the women in years	Ghudani Kalan Users No. & %	Ghudani Kalan Non-users No. & %	Ghudani Kalan Total	Grand Total Users No. & %	Grand Total Non-users No. & %	Grand Total Total
15—19	—	3 (100.)	3 (100.)	2 (9.52)	19 (90.48)	21 (2.60)
20—29	17 (24.29)	53 (75.71)	70 (23.33)	52 (25.12)	155 (74.88)	207 (25.62)
30—39	36 (46.75)	41 (53.25)	77 (25.67)	117 (53.67)	101 (46.33)	218 (26.98)
40+	24 (16.00)	126 (84.00)	15 (50.00)	91 (25.14)	271 (74.86)	362 (44.80)
Total	77 (25.67)	223 (74.33)	300 (100.)	262 (32.43)	546 (67.57)	808 (100.)

to use contraception in future was rather small. This, thus, indicates that comparatively the number of acceptors of contraception was small and among the non-acceptors of contraception not many intended to use contraception in future.

Socio-Economic Variable and Acceptance of Contraception

It is generally held that the use of contraception varies with the socio-economic status of the people and more acceptors belong to higher socio-economic background, as such, people generally have better knowledge and greater access to information regarding family planning. Table 11.5 gives the distribution of respondents by their own and their husband's educational level, their husband's occupational status and level of living among the contraceptors and the non-contraceptors.

Educational status of the individual is one of the important factors that influences the motivation to adopt contraception. Education, especially, higher education, enables a person to acquire better knowledge and also knowledge of more effective methods. The data in Table 11.5 reveal positive association between the educational attainments of the respondents and use of contraception in every village. Similar trend was observed with husband's educational status and, perhaps, this association was more reliable as the number of husbands who were educated upto various levels was greater than that of the wives. But these relationships were statistically significant at 0.05 level.

The data also revealed a close association between the occupational status of the husbands and use of contraception in each village in the sample and this relationship was also significant at 0.05 level. However, no significant relationship was found with the employment status of women.

The level of living of the respondents also indicated positive association between the two and this relationship was also significant at 0.05 level.

Thus, on the whole, the use of contraception was directly associated with the socio-economic characteristics of the respondents.

In the chapter on socio-economic differentials in fertility it was pointed out that in the longer duration of married life, i.e.,

TABLE 11.5 (a)

Distribution of respondent's by their own and their husbands educational status, husband's occupational status and level of living Vs. contraceptors and non-contraceptors

	Ugala			Ghagga		
	Users	Non-users	Total	Users	Non-users	Total
	(1)	(2)	(3)	(4)	(5)	(6)
Education of wife						
Illiterate	57 (32.20)	120 (67.80)	177	85 (35.42)	150 (64.58)	240
Below matriculation	19 (43.18)	25 (56.82)	44	9 (37.50)	15 (62.50)	24
Matric and above	9 (64.29)	5 (35.71)	14	6 (56.67)	3 (33.33)	9
Total	85 (36.17)	150 (63.83)	235 (100.00)	100 (36.63)	173 (63.37)	273 (100.00)

TABLE 11.5 (a)—*Contd.*

	(1)	(2)	(3)	(4)	(5)	(6)
Education of husband						
Illiterate	31 (29.52)	74 (70.48)	105	46 (32.39)	96 (67.61)	142
Below matriculation	29 (38.16)	47 (61.84)	76	32 (40.00)	48 (60.00)	80
Matric and above	25 (51.02)	24 (48.98)	49	22 (55.00)	18 (45.00)	40
Not recorded	—	5	5	—	11	11
Total	85	150	235	100	173	273
Occupational status						
1—3	9 (39.13)	4 (60.87)	13	6 (37.50)	10 (62.50)	16
4—5	46 (35.65)	83 (64.34)	119	60 (43.17)	79 (56.83)	139
6—7	29 (35.37)	53 (64.63)	82	33 (33.00)	67 (67.00)	100
No cocupation	1 (9.09)	10 (90.91)	11	1 (5.56)	17 (94.44)	18
Total	85 (36.17)	150 (63.83)	235 (100.00)	100 (36.63)	173 (63.37)	273 (100.00)

Level of living						
Low level	30 (33.33)	60 (66.67)	90	25 (30.86)	56 (69.14)	81
Medium level	19 (36.34)	33 (63.16)	52	34 (36.90)	58 (63.04)	92
Higher level	36 (38.71)	57 (61.29)	93	41 (41 00)	59 (59.00)	100
Total	85	150	235	100	173	273

TABLE 11.5 (b)

Distribution of respondent's by their own and their husbands educational status, husband's occupational status and level of living Vs. contraceptors and non-contraceptors

	Ghudani Kalan			Grand Total		
	Users	Non-users	Total	Users	Non-users	Total
	(1)	(2)	(3)	(4)	(5)	(6)
Education of wife						
Illiterate	48 (21.92)	171 (78.08)	219	190 (29.97)	446 (70.13)	636
Below matriculation	22 (38.60)	35 (61.40)	57	50 (40.00)	75 (60.00)	125
Matric and above	7 (29.17)	17 (70.83)	24	22 (46.81)	25 (53.19)	47
Total	77 (25.67)	223 (74.33)	300 (100.00)	262 (32.43)	546 (67.57)	808 (100.00)

Education of husbands						
Illiterate	36 (25.90)	103 (74.10)	139	113 (29.27)	273 (70.73)	386
Below matriculation	19 (27.54)	50 (72.46)	69	80 (35.56)	145 (64.44)	225
Matric and above	22 (37.93)	36 (62.07)	58	69 (49.64)	78 (53.06)	147
Not recorded	—	34	34	—	50	50
Total	77	223	300	262	546	808
Occupational status						
1—3	3 (50.00)	3 (50.00)	6	18 (40.00)	27 (60.00)	45
4—5	51 (30.54)	116 (69.46)	167	157 (36.94)	268 (63.06)	425
6—7	20 (31.25)	44 (68.75)	64	82 (33.33)	164 (66.67)	246
No occupation	3 (4.76)	60 (95.23)	63	5 (5.43)	87 (94.57)	92
Total	77 (25.67)	223 (74.33)	300 (100.)	262 (32.43)	546 (67.57)	808 (100.)

TABLE 11.5 (b)—*Contd.*

	(1)	(2)	(3)	(4)	(5)	(6)
Level of living						
Low level	11 (22.00)	39 (78.00)	50	66 (29.86)	155 (70.14)	221
Medium level	15 (23.81)	48 (76.19)	63	68 (32.85)	139 (67.15)	207
Higher level	51 (27.27)	136 (72.73)	187	128 (83.68)	252 (66.32)	380
Total	77	223	300	262	546	808

20 years and more non-migrants had higher number of live-births than the migrants, it was expected that the use of family planning methods would be more among migrants as compared to non-migrants. Majumdar and Dass (1972) in their study on some aspects of adoption of family planning practices by migrants and old settlers in five villages of West Bengal also revealed that the migrants had a more favourable attitude towards family planning practices than old settlers and also migrants adopted institutional sources of family planning more than the old settlers.

The data on the whole also revealed that the acceptors of contraceptors were more among migrants as compared to non-migrants although the difference between the two categories was small.

REFERENCES

1. Majumdar, A.K. and Das, K.D. 1972: "Some Aspects of Adoption of Family Planning Practices in Indian Villages", The Journal of Family Welfare, 18, (3) March.
2. Rele, J.R. and Patankar, T. 1969: Differential Fertility of Contraceptors and Non-contraceptors. A Paper presented in a general conference by IUSSP, London.
3. Operation Research Group 1972: An all India Survey of Family Planning Practices, Baroda, India.
4. Gandotra, M.M. and Narayan Dass 1981: "Fertility Differentials amongst Contraceptors and Non-contraceptors in India", Demography India, Nos. 1 & 2.
5. Vig, O.P. 1981: "Family Planning Practices and its Impact on Fertility in Maharashtra", Demography India, Nos. 1 & 2.

12

SUMMARY AND CONCLUSIONS

THE present study was undertaken to analyse the present levels of fertility in the rural communities of Punjab and Haryana and examine some of the socio-economic and sociological correlates of fertility behaviour prevalent in these rural areas. Besides these inter-village variations, an attempt has also been made to examine inter-village variations in fertility behaviour, if any, in these rural communities.

Although there are large number of studies dealing with fertility differentials but most of these studies taken in the consideration socio-economic variables only. The distinctive feature of the present study is that it takes into account some of the important sociological variables, like caste status, family structure, religion and level of religiosity and modernity value orientations of the women and provide explanations for observed fertility differentials, thus incorporation sociological perspective. Another important reason for taking up the present study was that there were hardly any studies of fertility differentials in these states, especially, pertaining to rural areas, which had been undergoing rapid economic transformation due to green revolution.

The study was conducted in three villages, namely, Ugala, Ghagga and Ghudani Kalan, chosen from the states of Punjab and Haryana which had been undergoing rapid economic changes in recent decades. The villages were fairly large in size and have predominently agricultural economics. They were quite free from urban influence being far away from the nearby main cities and had consistently very high (Ugala and Ghagga) or very low (Ghudani Kalan) rates of population growth during the past three decades. Simple random sampling procedure was adopted, and accordingly, 33 per cent of the households were included in the study. The total number of women interviewed came out to be 808.

The analysis of the data shows that the average age at marriage for the total sample population was 17.80 years with no significant inter-village variations which is indicative of the fact that rural communities in Punjab and Haryana have more or less the same pattern of age at marriage. The data further shows that the younger women below the age of 30 had comparatively higher age at marriage showing thereby, the rising trend in the age at marriage among younger women.

Age at marriage was found to be positively related to the socio-economic status of the respondents and negatively related to the number of live-births and this relationship was more significant when controlled for marriage duration in every village. Some inter-village variations were discovered though they were not very significant.

The analysis of fertility behaviour reveals that although 4.53 was the mean number of live-births for the total sample population, however, it was 4.82, 4.73 and 4.12 live-births, respectively, for Ugala, Ghagga and Ghudani Kalan. The trends in general mortality were just opposite to that of fertility as mean number of deaths per woman for the total population was 0.76, whereas, it was 0.68, 0.72 and 0.83 for Ugala, Ghagga and Ghudani Kalan, respectively. Pregnancy wastage was reported to be quite low in these villages.

The number of male-births was, on the whole, higher than the female births in all the three villages and the difference between Ugala and Ghagga was small but quite substantial for Ghudani Kalan. In terms of parity more women in Ugala had higher parity births, i.e., more than four live-births and comparatively the proportion of such women was much smaller in Ghudani Kalan while Ghagga came at the middle level.

However, taking into account the present average age of the respondents (38.12 years, 37.82 and 38.66 years in Ugala, Ghagga and Ghudani Kalan, respectively) and the total duration of married life (20.47 years, 19.48 and 20.16 years in Ugala, Ghagga and Ghudani Kalan, respectively) the present level of fertility was found to be higher in Ugala and lower in Ghudani Kalan while Ghagga falls at the intermediate level.

The trends in general mortality were just contrary to fertility as its incidence was comparatively higher for Ghudani Kalan aud lower for Ugala, while Ghagga again fell at the intermediate level. On the other hand the incidence of infant and child deaths was higher for Ghagga as compared to the other two villages. Age and sex differentials in infant and child mortality were quite significant for all the villages and majority of the deaths occurred during the neo-natal period, followed by post neo-natal stage and the proportion of deaths in the age group of 1-4 years was comparatively much smaller. Another notable point was that there were more male deaths during the neo-natal period, whereas, there were more female death during the post neo-natal and later age groups.

The analysis of data revealed significant correlation between infant and child mortality and fertility as the higher the number of births, the higher was the number of deaths and vice-versa for all these villages. Incidence of infant and child mortality also varied with the birth order and age of the mother.

The extent of infant and child mortality was also found to vary with the socio-economic status of the respondents. Nevertheless, given the same level of the socio-economic status of the respondents, the extent of infant mortality revealed significant inter-village variations, being invariably higher for Ghudani Kalan and lower for Ugala.

Among socio-economic variables, educational level of the respondents and their husbands was found to be negatively related to the number of live-births although the number of educated respondents and their husbands was small. Number of births was comparatively small not only among gainfully employed women but it also varied with the employment status of the women being much smaller among women with higher education and employed in prestigeous occupations as against agricultural workers. But the occupational status of the husband did not reveal any correlation with the number of births, may be because these were mostly rural occupations not associated with any higher technical education. Likewise family income did not show any correlation with the number of births. However, level of living which is a combination of various indicators, revealed negative correlation with fertility in all the villages.

Since the study pertained to the rural areas and more than half of the respondents came from the families of land owners, it was significant to note that land owners had smaller number of births as compared with those who did not own any land. Further, among the land owners, big farmers (those who owned 25 or more acres of land) had smaller number of births as compared to medium and small farmers for all the three villages.

Place of dwelling revealed negative relationship to the number of births in the total sample population although there was no clear trend for each village. Lastly, though migrants on the whole, had higher number of births than non-migrants, but in the longer duration of marriage (20 years and more) where women had either completed their reproductive span, or, they were near its completion, non-migrants had higher number of births.

It may also be mentioned here that given all these socio-economic variables, number of births with few exceptions was higher for Ugala and lower for Ghudani Kalan.

The caste status of the respondents indicated negative relationship with the number of births in the total sample as well as in village Ghudani Kalan, and, with one exception, in Ghagga, but no consistent trend was observed in Ugala. This relationship was found to be more significant in the longer duration of marriage where women has been married for more than 20 years. The significant point noted was that inter-village variations in the number of births were quite substantial among higher and middle caste respondents but they were very small among lower caste women which showed that backward and scheduled caste women had more or less the same level of fertility in all these villages.

The analysis of data further indicated that backward and scheduled caste respondents, on the one hand had, higher number of births, and, on the other, they had comparatively lower age at marriage, large majority of them were either illiterate or had low level of education, majority of their husbands were employed in low prestige jobs and had comparatively low family income, therefore, these variatons in the socio-economic and demographic variables could account for their higher number of births.

So far as religion is concerned, no uniform trend was noticed in all the villages as number of live-births was higher among the Hindu respondents in Ugala and Ghagga but opposite was the case in Ghudani Kalan. But in the longer duration of marriage, i.e., 20 years and more, the Muslim respondents (though small in number) had the highest number of births, followed by the Hindus and the Sikh respondents had comparatively smaller number of births in all the three villages and the differences in these three groups were quite significant. It was further noted that in the two villages where, on the whole, the Hindu respondents had higher number of births, they also had comparatively lower age at marriage and higher incidence of infant and child deaths and opposite was true for the Sikh respondents in Ghudani Kalan where they had higher number of births. Nevertheless, there was no significant difference in the socio-economic status of these two major religious communities although the level of living was comparatively higher among the Sikh respondents.

Level of religiosity of the respondents did not show any significant correlation, although, on the whole, the number of births were higher among highly religious respondents.

Coming to the modernity orientations of the respondents, the data revealed that modern respondents had smaller number of births as against the traditional respondents and this trend was further confirmed when controlled for marriage duration. At the same time, they also had comparatively higher age at marriage, lower incidence of child loss, greater interspouse communication and greater use of contraception and these factors could, therefore, account for their lower level of fertility.

Among familial variables, the first important variable was the family structure. The findings show that although on the whole, in all these villages, the number of live-births were higher in the nuclear than in, the joint families, but in the longer duration of marriage, i.e., 20 years and more, number of births were higher in the joint as against the nuclear families. But in both the cases the differentials were weak and the intermediate variables like age at marriage or extent of child loss and contraception use did not reveal any consistent trend.

The level of interspouse communication revealed negative relationship with the number of births. It was observed that the use of contraception was much higher among those respondents who had discussed the number of children they would like to have and the use of contraception and the higher was the use of contraception, the smaller was the family size.

Another familial variable was the desired ideal family size. Large majority of the respondents in all the three villages preferred three to four children as an ideal family size and expressed greater preference for sons than daughters although a daughter was also considered necessary for a family. The preference for an ideal family size was related to the respondent's age and socio-economic status.

Children are valuable to parents as they provide social, economic and psychological satisfaction but parents generally do not attach similar values to children of both the sexes and the findings of the present study also indicate that the majority of the respondents had different values or reasons for having sons and daughters. The sons are desired primarily for their role in continuing the family name or for the provision of old age security and financial aid to the family whereas daughters are desired for Kanyadan or domestic help they render to the family although some of them even mentioned that daughters when married increased the network of familial relations which is always beneficial to the family.

Lastly, in order to gauge the impact of family planning practices on fertility behaviour, it was noted that the proportion of adoptors of contraception was rather small as only one third of the women were ever-users of contraception. Further, of this number, two thirds of them (65 per cent) had adopted permanent methods and only one third of them were using other methods. Since majority of these users of contraception adopted it for limiting the family size rather than for spacing, the adoption was generally after 4th or 5th pregnancy, as a result, the adoptors of contraception had higher number of births as compared to the non-adoptors. Further, number of births varied not only with the type of methods used but also by the number of surviving sons.

Besides these inter-village variations in the fertility behaviour of the women, another significant finding of the study is the inter-village variations in the fertility level of the respondents. In other words given the same level of socio-economic status and other sociological variables, fertility level was invariably found to be higher in Ugala and lower in Ghudani Kalan while Ghagga fell at the intermediate level, irrespective of the fact that there were no significant variations in the age at marriage, duration of marriage and present average age of the respondents.

However, these inter-village variations could be explained in terms of the soico-economic and structural aspects of these villages. For example, of the three villages under study, Ghudani Kalan (District Ludhiana) was more developed in terms of socio-economic transformation as compared to the other two villages. The village was more developed in service sector and it had greater educational and health care facilities. On the other hand, Ugala (District Ambala) was the least developed village not only in the economic sphere but also in terms of educational and health care facilities. Like Ghudani Kalan, Ghagga (District Patiala) was also more highly developed than Ugala but in terms of trade and commerce, communication and administration. Further, the socio-economic development of these villages seems to be reflected in the socio-economic background of the respondents as it was observed that in terms of the socio-economic status of the respondents, Ghudani Kalan was better than the other two villages and it was further confirmed by the fact that level of living of the people in this village was, on the whole, much higher than in the other two villages.

Caste, which is an important aspect of social structure of the communities under study, also seems to be an important factor in causing these differentials. It has been observed that the castes on the lowest rung of the status hierarchy have generally large families and Ugala had the distinction of having the highest proportion of backward and scheduled caste respondents.

Thus, on the whole, some of the variations in the socio-economic development of these villages resulting in differentiation in the socio-economic status of these respondents might had contributed to the variations in the fertility level, among these villages.

Conclusions

The conclusion, therefore, is that all the variables discussed above had been responsible for differentially influencing the fertility behaviour of rural women in these villages. Among all these variables, one of the important findings of the study is that land owners have lesser number of births than those who did not own any land. Similar results were also reported in the Khanna Study (Wyon and Gordon 1971). In that study also the fertility of cultivating castes was found to be substantially lower than that of other higher castes of Brahmins, Banias, Khatris, on the one hand and that of other lower castes on the other. In a study conducted in Tamil Nadu, a region which is very different from Punjab and Haryana in its culture and geography, it was also found that the major landowning caste of Vellalas have the smallest family size among the people of the region and the trading communities of Muslims had the largest family size (Umran and Standley 1976). As pointed out by D'Souza (1983), it seems that landowning and farming castes not only in Punjab and Haryana but all over India are experiencing the conditions of downward mobility and hence having relatively smaller family size

In addition to the above mentioned variables certain sociological variables which have hitherto been neglected by most of the studies also reveal significant influence on fertility behaviour although these variables do not influence fertility behaviour per se. Their influence is observed through certain intermediate variables like age at marriage, extent of child loss, interspouse communication, adoption of contraception and the socio-economic background of the women.

Thus, the appearance of these differentials in the level of fertility enables us to observe that fertility decline has started in these rural communities although the process is slow.

The important question that arises now is whether we can

say on the basis of these variations, that fertility differentials have emerged in these rural communities as a result of variations in the socio-economic conditions, in the region, as a whole. Since the villages under study, though selected from different districts, had more or less the same characteristics, in spite of some small variations, it should be possible for us to generalize for the rural communities of the region as a whole, on the basis of the findings of this study.

However, some variations among the villages could be explained in terms of the level of socio-economic development of these villages which suggested that with greater socio-economic development these differentials would widen and more and more people would reduce the size of their family. Of course, this does not mean that economic development is the key to differential fertility but it is one of the important factors to be taken care of for bringing down the level of fertility in rural areas. Moreover, identification of these variables on the basis of variations in fertility behaviour would enable the planners to rationalize the inputs of the family planning programme.

BIBLIOGRAPHY

Agarwala S.N., *Age at Marriage in India,* Bombay, Kitab Mahal.

Ibid., "Social and Cultural Factors affecting Fertility in India", *Population Review*, 8.

Ibid., *India's Population Problem,* New Delhi, Tata McGraw Hill Publishing Company.

Bebarta, P.C., "Attitude of Women towards Family Planning —A Study of Differences by Family Type in Six Villages of Delhi", *Quarterly Journal of Indian Studies in Social Sciences*, 1(1).

Berelson, Bernard, "Social Science Research on Population", *Population and Development Review*, Vol. 2.

Bongaarts, John, "A Framework for Analyzing the Proximate Determinants of Fertility", *Population and Development Review*, Vol. 4, 1.

Bulsara, Jal F., *Problems of Rapid Urbanization in India*, Bombay, Popular Prakashan.

Census Commissioner India, *Census of India*, 1971, Social and Cultural Tables, Series I, Part II-C(i), Office of the Registrar General.

Ibid., *Census of India*, 1971, Social and Cultural Tables, Series I, Part II-C (i), Office of the Registrar General.

Ibid., *Census of India*, Series 1, Paper No. 3 of 1981.

Chandrasekharan, C. and George, M.V., "Mechanisms Underlying the Differences in Fertility Patterns of Bengalee Women from Three Socio-Economic Groups", *Milbank Memorial Fund Quarterly*, 40.

Chandrasekhar, S., *Infant Mortality, Population Growth and Family Planning in India*, London, George Allen and Unwin.

Chaudhury, R.H., *Social Aspects of Fertility*, New Delhi, Vikas Publishing House.

Clifford, William, B., "Modern and Traditional Value Orientations and Fertility Behaviour: A Social Demographic Study", *Demography*, Vol. 8, 1, Feb. 1971.

Das Gupta and Others, *Couple Fertility*, NSS Report, 7.

Davis and Blake, "Social Structure and Fertility : An Analytical Framework", *Economic Development and Cultural Change*, Vol. 4.

Davis, K., *The Population of India and Pakistan*, Princeton, Princeton University Press.

Ibid., *Population Policy : Will Current Programmes Succeed*? San Francisco, Population Evaluation and Birth Control, San Francisco, Garrett Harden, Freeman and Company.

Ibid., *Population Policy : Will Current Programmes Succeed*? Science, C1 viii.

Devi, V.S., "Effective Perception of Infant Mortality on Actual Family Size", *The Journal of Family Welfare*, Vol. xxiv, 4.

Driver, E.D., "Fertility Differentials among Economic Strata in Central India", *Eugenic Quarterly*, 7, (2) June.

Ibid., *Differential Fertility in Central India*, *Princeton*, Princeton University Press.

El-Badry, M.A., "A Study of Differential Fertility in Bombay", *Demography*, 4.

Freedman, Ronald, "The Sociology of Human Fertility: A Trend Report and Bibliography", *Current Sociology*, Vol. X/XI, No. 2.

Ibid., "Application of the Behavioural Sciences to Family Planning", *Studies in Family Planning*, 23, October.

Gandotra, M.M. and Narayan Dass, "Fertility Differentials amongst Contraceptors and Non-contraceptors in India", *Demography India*, Nos. 1 and 2.

George, Molly Kutty, "Application of Mahalanbois D^2 in a Study of Tradition and Fertility", *The Journal of Family Welfare*, Vol. 19. 1.

Ghosh, B., "Policy Oriented Research into Determinants of Fertility", *Demography India*, Vols. 1 and 2.

Glass, D.V. and Grebcnick, "The Trend and Patterns of Fertility in Great Britain", *Papers of the Royal Commission on Population*, London, HMSO 2 Vols,

Govt. of India, *Sample Registration Bulletin*, Vol. xiii, 2 (Vital Rates) December.

Govt. of India, *Fertility Differentials in India*, Office of the Registrar General and Census Commissioner, New Delhi.

Hawthorn, Geofrey, *The Sociology of Fertility*, London, Collier Macmillan.

Goyal, R.P., "Shifts in Age at marriage in India between 1961-71", *Demography India*, Vol. iv, 2.

Heer, D.M., "Economic Development and Fertility", *Demography*, 3.

Hill, R., Stycos, M.J. and Back, K.W., *The Family and Population Control: A Puerto Rican Experiment in Social Change*, Chapel Hill, University of North Carolina Press.

Hussain, I.Z., "Educational Status and Differential Fertiliy in India", *Social Biology*, Vol. 17.

Jagdish, C. Bhatia and George Sabagh, "Migration and Fertility in India", *Demography India*, Vol. 9, Nos. 1 and 2, Jan-Dec.

Jain, S.P., *Demography—A Status Study on Population Research in India*, Vol. II, New Delhi, Tata McGraw-Hill.

Ibid., "Levels and Differentials of Infant and Child Mortality: Determinants and Demographic Impact", in Srinivasan, K. et al. (eds.) *Demographic and Socio-Economic Aspects of the Child in India, Bombay,* Himalaya Publication House.

Karkal, M., "The Family and Fertility", *The Journal of Famliy Welfare*, Vol. XXII, 1.

Kiser, Clyde V. and Whelpton, P.K., "Resume of the Indianapolis Study of Social and Psychological Factors Affecting Fertility", *Population Studies*, Vol. 7.

Lorimer, F. et al., *Culture and Human Fertility*, Paris, Unesco.

Mhadevan, K., *Sociology of Fertility: Determinants of Fertility Differentials in South India*, New Delhi, Sterling.

Majumdar, A.K. and Das, K.D., "Some Aspects of Adoption of Family Planning Practices in Indian Villages", *The Journal of Family Welfare*, 18, (3) March.

Mamdani, M., *The Myth of Population Control*, London, Monthly Review Press.

Mitra, Asok, *India's Population: Aspects of Quality and Control*, Vols. I and II, New Delhi, Abhinav Publications.

Mukherjee, B.N., Prediction of Family Planning and Family Size from Modernity Value Orientations of Indian Women, *Papers of the East-West Population Institute*, Hawaii, No. 61

Nam, Charles B., *Population and Society*, Boston, Miffin.

Nayar, P.K.B., "The Influence of Education on Fertility", *The Journal of Family Welfare*, Vol. XX, 3.

Omran, A.R. and C.C. Standley, *Family Formation Patterns and Health*, Geneva: World Health Organization.

Operation Research Group, *An all India Survey of Family Planning Practices*, Baroda, India.

Pakrasi, K. and Malakar, C., 'The Relationship between Family Type and Fertility', *Milbank Memorial Fund Quarterly*, 45, (4).

Pareek, U. and V. Kothandapani, "Modernization and Attitude towards Family Size and Family Planning: Analysis of Some Data from India", *Social Biology*, Vol. 16, 1.

Pareek, U. and Rao, T.V., *A Status Study on Population Research in India*, Vol. I, New Delhi, Tata McGraw Hill.

Poffenberger, Thomas, *Husband-Wife Communication in an Indian Village*, New Delhi, Central Family Planning Institute.

Potts, Malcolm and Selman, Peter, *Society and Fertility*, London, Macdonald and Evans.

Rainwater, Lee, *And the Poor Get Children: Sex Contraception and Family Planning in the Working Class*, Chicago: Quadrangle Book.

Reddy, P.H. and Gopal Y.S., "Socio-Economic Factors and Age at Marriage", *The Journal of Family Welfare*, Vol. xxvi, 3.

Rele, J.R., "Fertility Differentials in India—Evidence from a Rural Background", *Milbank Memorial Fund Quarterly*, Vol. XII, 2, April.

Rele, J.R., and Patankar, T., "Differential Fertility of Contraceptors and Non-contraceptors", *A Paper Presented in* a General Conference by IUSSP, London.

Rele, J.R. and Kanitkar, "Fertility Differentials by Religion in Greater Bombay: Role of Explanatory Variables", in *The Economic and Social Supports for High Fertility* (ed.) L.T. Ruzicka, Canberra, Australian National University.

Ruzicka, L.T. and Kanitkar, Tata, "Infant Mortality in an Urban Setting: The Case of Greater Bombay", in Vaidyanathan (ed.), *Studies on Mortality in India*, Ghandigram, Institute of Rural Health and Family Planning.

Saxena, G.B., "Differential Fertility in Rural Hindu Community: A Sample Survey of Rural Uttar Pradesh, India", *Eugenics Quarterly*, 12, (3) Sept.

Singh, K.P., *Status of Women and Population Growth in India*, New Delhi, Munshiram Manoharlal Publishers.

Singh, K.P., "Modernity Value Orientations, Fertility and Family Planning", *The Journal of Family Welfare*, Vol. 129, 2.

Singha, P., "Infant Mortality and the Level of Fertility in India", A Review, *Demography India*, Vol. 4, 2.

Srinivas, M.M. and Ramaswamy, E.A., *Culture and Human Fertility in India*, New Delhi, Oxford University Press.

Stockes, C.S., "Family Structure and Socio-Economic Differentials in Fertility", *Population Studies*, 27 (2).

Stycos, J.M., "Obstacles to Programmes of Population Control: Facts and Fancies", *Marriage and Family Living*, 25.

Thompson, W.S. and Lewis, D.T., *Population Problems*, New York, McGraw Hill Publishers.

United Nations, *Mysore Population Study*, New York.

Ibid., The Determinants and Consequences of Population Trends, New Summary of Findings on Interaction of Demographic, Economic and Social Factors, *Population Studies*, No. 50, New York.

Ibid., *Husband-Wife Communication and Practice of Family Planning*, Bangkok, Asian Population Studies Series, 16.

Ibid., Demographic Transition and Socio-Economic Development, Istambul, *Proceedings of the United Nations Expert Group Meeting.*

Vasaria, Lella, "Religious Differentials in Fertility", in *Population in India's Development from 1947 to 2000* (ed.) Ashish Bose, Mitra, New Delhi, Vikas Publishing House.

Victor, S. D'Souza, *Economic Development, Social Structure and Population Growth*, Population Research Centre, Department of Sociology, Punjab University, Chandigarh (Mimeographed).

Vig, O.P., "Family Planning Practices and its Impact on Fertility in Maharashtra", *Demography India*, Nos. 1 and 2.

Wastoff, C.F. et al., *The Third Child: A Study in the Prediction of Fertility*, New Jersey, Princeton University Press.

Wyon, and Gordon J.E., *The Khanna Study: Population Problem in the Rural Punjab*, Cambridge, Harward University Press.

INDEX